# *How many times had he fantasized about this moment?*

In how many of them had he found her and killed her, found her and loved her, found her and killed himself. And now—now to find she'd been with him all this time, with him totally ignorant of whom she really was.

The irony of it burned him. If not for what had happened the night before, he might have never known. A rough, barked laugh escaped him. He would have never known.

Almost against his will, his head turned so he could look over his shoulder. There she stood, looking concerned and innocent, the catalyst of his current life, the bane of his existence. A maelstrom of emotions swept through him, threatening to drive him mad. Joy, anger, love, hate, triumph, loss—how could he feel these things all at once?

"Torren?" Her arms about herself, looking hurt and confused, Larana took a tentative step toward him.

"Stay back!" He glared at her, his body shaking, a war he'd not expected waging inside him.

# ALSO BY GLORIA OLIVER

*In Service of the Samurai*
*Willing Sacrifice*

# VASSAL

## OF

## EL

# GLORIA OLIVER

ZUMAYA OTHERWORLDS                          AUSTIN TX

2004

This book is a work of fiction. Names, characters, places and incidents are products of the author's imagination or are used fictitiously.

VASSAL OF EL

© 2004, 2008 by Gloria Oliver

ISBN 978-1-934135-99-0

Cover design © Martine Jardin

First published by Zumaya Publications 2004

"Zumaya Otherworlds" and the griffon logo are trademarks of Zumaya Publications LLC, 3209 S. Interstate 35, Austin TX. Look for us online at http://www.zumayaotherworlds.com

Library of Congress Cataloging-in-Publication Data

Oliver, Gloria
  Vassal of El / Gloria Oliver.
      p. cm.
  ISBN 978-1-934135-99-0 (alk. paper)
  1.  Soldiers of fortune--Fiction. 2.  Young women--Fiction. 3.
Attempted murder--Fiction. I. Title.

PS3615.L5845V37 2008
813'.6--dc22

                           2008046985

# ACKNOWLEDGMENTS

Lots of thanks to my editors Cindy Speer and Elizabeth Burton, as well as Ivy Reisner for their help with this book, but special mucho thanks to Wendy Dinsmore for her excitement and great guidance. This book would not be what it is without them.

# CHAPTER 1

*R*ED, EVERYTHING WAS RED. IT OOZED AND DRIPPED AND *covered everything. It pressed down over him, stifling him. Moans rang all around. There were shrieks of pain and then silence. Suddenly there were hands, dozens of them, grabbing, pinching, lifting him...*

With a gasp, Torren sat up in the tree-deepened darkness. He studied the forest around him, the feeling of those hands still with him as his breath rushed wildly in and out. A few burning embers in the small pit of his fire twinkled back at him, showing him he was alone.

His pack and boots still leaned against the large maple closest to him. His sword was in its scabbard at his side, as he'd left it, within easy reach. His breathing slowed, these facts, one-by-one, calming him. The perspiration on his sun-weathered face and arms turned cold and made him shiver. The last tendrils of the dream left him.

Wiping his face and close-cropped platinum hair, he flicked his blanket aside and climbed to his bare feet. Chiding himself as he took yet another look around to make sure he was alone, he hobbled over to his pack to change his sodden shirt.

The dreams—the memories—hadn't troubled him in more than a year. He'd actually started to hope they were gone for good. They were an annoyance, and a horrid reminder of things he'd rather forget. Not, it seemed, that he ever could.

He wasn't sure if it was because he'd not suffered the

dream in a long time; but this time it had felt sharper, more immediate than ever before. He thrust the thought aside, not wanting to look at it too closely.

Pulling his sodden shirt over his head, he shuddered as the night breeze touched his skin. Setting the garment out to dry over a low branch, he quickly retrieved another out of his pack and slipped it on. He sat down on his damp blankets with a grimace, not sure he could sleep again, and glanced up through the overhanging branches of the maple at the sky.

Two of the three moons were still visible. He sighed, figuring he still had about six hours left before dawn. To-morrow, if he were lucky, he would run across a farm or other travelers on the road and possibly hear more about the happenings up toward the northern border.

A range of mountains stood on the boundary between the empire and Galt. For generations, it had helped main-tain an uneasy peace between the two, the trouble of mov-ing massive armies over the few passes carved through them, and the likelihood of ambush while doing so, too risky to make it worthwhile.

Recently, though, it seemed matters had changed. Whispers of war were in the air. Forces were supposedly gathering near the border. Weapon sales had increased. If half of what he'd heard so far was true, there was a good chance he'd be able to offer his services in the area as a bodyguard or mercenary and perhaps get a better-than-average wage.

Torren frowned as a slow shadow crossed the smaller of the two moons. He stared at it, the long mass cutting across the bright surface as it drifted through the sky—one of the floating cities of El. The moonlight gleamed off some of its tall spires, making them appear like jewels. The protective field over the island shimmered like star-dust.

The aerial cities of El—home to His people, the Chosen. A culture apart, living on their islands and high reaches where no mere mortals tread. A fantasy paradise, if you believed what half of those living on the ground adhered to, though in truth no Lander had ever been within the floating cities.

The empire still spent inordinate amounts of money trying to figure out how to tap the magic that kept the islands aloft and shielded them from the weather. Others tried to worship El instead, since he had supposedly created the islands for his people as gifts before being closed out of the world like the other gods by the First Mother. Neither method had yet to bear any fruit.

He was sure the other empires of the world were probably doing much the same. Though the Shirak Empire had little contact with those across the wide oceans, the Chosen did. Somehow, he doubted those other countries' feelings about them could be too far from those felt here. As far apart as the continents lay, and as treacherous as the waters were, gaining the secret of the islands or their flying ships would be a boon to whomever could replicate them. Then the Chosen would not be the only ones linking the world with commerce.

With a snort, he lay back down and turned from the sight. First the dream and now this—would he never be free of them? He had no need of those places, nor of their inhabitants. Yet, though he'd turned away from the drifting islands, he could still feel the pull of a Chosen city as it traveled across the sky, almost as if it were calling to him. He closed his eyes, trying his best to shut the feeling out, with little success. The island's presence, the fact he'd had the dream again and knowing he was close to the area where—

Torren stiffened, dropping his train of thought as a faint rustling came from somewhere behind him. He sat up and turned, automatically reaching for his scabbard. He'd half-risen, partially drawing his blade, when someone burst from the darkness and plowed into him in a tangle of arms and legs.

He fell back and, using the momentum, grabbed the intruder and flipped him to the ground, pinning him under his weight. He was slightly taken aback as he looked down at the face of his attacker in the dim light and saw a frightened girl.

"What do you think you're doing?" he demanded gruffly.

Wide eyes stared at him unseeing as she struggled in vain to get out from under him. Her breath came in harsh

3

gasps, her arms and face were scratched and bleeding from running through the brush.

He kept her pinned, wondering what someone like her was doing out here at this time of night. He assumed she came from a nearby farm. Unfortunately, the fire had burned too low to see anything clearly aside from her gender.

Slowly, the girl's struggles subsided; she focused on him for the first time. Tears welled in her large eyes as they locked with his.

"Help me. Please, help me."

He released her, sitting beside her in an attempt to keep her calm. "Are you being chased?"

She sat up, a shudder running through her as she wrapped her arms around herself. She nodded. "I lost them, I think. But..."

Picking up his sword and scabbard, he strapped them on as he stood, staring in the direction she'd come from. If she was being followed, her pursuers couldn't be far behind. There were no sounds heralding their arrival yet, but he knew he'd still have to work quickly if he were to be prepared. Anyone about at this time of night couldn't be up to much good.

"Grab the blankets and go stand by that tree." He pointed over to the maple his pack was leaning against. Turning away, he put on his boots and then, with his foot, quickly pushed dirt over the fire pit. With a faint hiss and rising smell of ashes, the embers were buried, what little light they'd been emitting gone. He then shoved leaves over the newly covered hole and waved at the air around him to dispel any of the remaining smoke.

Still keeping an ear tuned to the muted night sounds around him, he hurried back to the large tree where the girl stood waiting for him. Cloaked in shadow, she was huddled against the massive trunk, holding the blankets she'd retrieved like a shield against her chest.

"Come on," he told her, "time to climb up."

The girl, who barely reached to his shoulder, only stared blankly at him.

Torren frowned. "We need to climb this tree. You don't want to be found, right?"

4

She shook her head rapidly from side to side yet made no move to do as he'd asked. Trying hard not to let his irritation show on his face, he yanked his damp shirt off the tree limb and shoved it into his pack. Turning to look at her again, he slung the pack onto his shoulders. She hadn't moved, still staring at him intently with her large eyes. He sighed silently—he'd have to do this the hard way.

He grabbed her by her small waist, eliciting a surprised gasp. He ignored it, lifting her. She gasped again and let go of the blankets, raining them down like leaves on his head.

Trying not to become even more annoyed than he was already, he spoke to her again. "Grab a limb and climb up. Do it now!"

Feeling her finally obeying, he let go of her, pulled the blankets off his head and settled them on his shoulder. After a moment, he climbed after her. The scent of the tree's bark was strong.

"Keep going. You need to get up into the thickest part of the canopy."

Without a word, the girl scurried into the higher branches without much trouble. The leaves barely rustled as she passed.

"That's far enough." He was forced to reach out and grab her by the ankle, since it looked as if she would keep going until she reached the top of the tree and beyond. "Sit there."

Timidly, she drifted back down and nestled where he pointed. Three limbs jutted out from a thick central branch, making a seat Torren hoped she'd have a hard time falling out of.

Making sure she was secure and looked to be staying put, he found a place for himself.

"Here, cover yourself with this. It'll make you harder to see." He handed over one of the blankets, though he already had trouble making her out amidst the foliage. The girl quickly wrapped herself in it, her teeth chattering softly.

Shaking his head as he watched her, he took the second blanket and wrapped up in it. When he was done, he ig-

5

nored her, instead concentrating his senses on the wooded landscape below. If the girl was being chased, her pursuers were late. With luck, she'd lost them in the brush, but it was best to make sure.

After several long minutes, the crickets, which had grown silent at her abrupt arrival and later started up again, went suddenly silent once more. A curse echoed through the small clearing.

"Why would she have come this way?"

Torren stiffened at the sound of the annoyed voice, not having sensed the stranger's presence until then. He glanced over at the girl and saw her duck her head inside the blanket in fear.

"She's stupid? How should I know?" said a second voice, sounding even more annoyed than the first.

"She won't be running through all this for long, though, that's for sure." The first one snorted. "Never seen anyone run so fast."

"Fear's a great motivator." The second man paused. "I think she may have gone this way."

The two men drifted closer. More curses colored the night as they were forced to deal with the brush.

Torren silently removed his sword from its sheath, then the large knife hidden in his boot. He considered giving the girl the dagger in his pack but rejected the idea. She was more likely to hurt *herself* with it than them.

The two men shoved their way out of the bushes into the small clearing and stopped. He watched them, not able to make out much in the dark. One scrunched closer to the ground.

"Which way?" asked the other.

The first was silent for almost a full minute as he tried to study the ground around him. "It's too dark. The signs aren't clear."

"Dek is *not* going to want to hear this."

The first snorted. "You don't know how lucky we've been to be able to follow her this far."

The other grunted in reply, not at all happy. "What now?"

Torren tensed.

The first rose to his feet. "We go back. What else? If

6

Dek still wants to find her, we can try to pick up her trail in the morning."

"So much for this being an easy job." The two men started back the way they'd come.

Torren slowly let himself relax. The fact they'd been able to track her at all from the road at night meant they were good. If they'd brought a light with them they would have surely been able to tell where she'd gone, and he'd have had no choice but to fight them.

The girl was likely a farmer's daughter—the closest town was a few days away—so why would people of such skill be after her?

He shook his head. It didn't matter. What did was that they'd be back. Once they examined the area in daylight, they'd realize the girl had run across someone. This would change the rules of the game. Depending on why they wanted her, they might decide to take offense at the fact he'd helped her. The more distance he could put between himself and these men before they came back the better.

"It's time to go."

The covered lump that was his unexpected companion didn't move. For a long moment, Torren considered just leaving her there. He knew he wouldn't, but he considered it, all the same.

With an irritated sigh, he got off his perch and reached over to remove the blanket from her head. As he threw the corner of it off her, she jumped in her seat with a small squeal.

"If you don't want them to find you, we have to go. Now." He pointed to the ground; and after a moment, she scampered down away from him with wide eyes. Though she wasn't what he'd call graceful, it looked as if she'd had experience climbing trees.

Following at a more sedate pace, he descended, going over their options. Traveling through the trees at night would be difficult, and he didn't know of any convenient streams nearby they could use to hide their passage. If he wanted to get away, it seemed they had no choice but to use the road. There would be nothing there to trip over, and the packed surface should hide any traces of their passing. Even better, he would use the pursuers' own trail to get back to the road to make things even more difficult for them in the morning—that would work just fine.

Torren glanced over at where the girl stood waiting for him, still huddled in the blanket. "Stick close to me. We're going to make our way back to the road."

She stiffened, her face looking wan in the moonlight. "No..."

His brow went up. "Suit yourself. You can stay here if you want. But they'll definitely find you in the morning."

He shrugged when she said nothing and started on his way, not caring one way or the other. If she didn't want his help, so be it. He hadn't gone far before he heard her struggling to catch up.

In less than ten minutes, they were at the road. Though not one of the empire's stone-paved highways, it was broad and followed a well-used route. Before stepping onto it, he glanced up and down to make sure the girl's two pursuers were nowhere near. Spotting no one, he left the shelter of the trees and started north. The girl left the concealment of the trees a minute or so later and followed.

Shadows played in the moonlit darkness to either side; but Torren ignored them, keeping his senses primed for living threats. They traveled for more than an hour and saw nothing and no one. Figuring he'd gone far enough to distance them from his old camp, he stopped and waited for the girl. He watched as she came up and almost bumped into him, stooped as she worked at putting one foot in front of the other.

"We're getting off," he informed her.

The spot he'd chosen was bare of bushes or small plants, and the surface looked to be hard enough they wouldn't leave much of a trail. Unless her pursuers had brought sniffers with them, which he doubted, they'd be hard-pressed to find where their quarry had abandoned the road.

"Step where I step."

He stared hard at the ground, trying to choose their path carefully. He avoided plants or areas of soft earth, for a cracked branch or indentation would give them away to anyone with tracking skill.

When he felt they'd gone far enough away from the

road, he searched for a place to stop. Finding a likely spot, he gratefully let his pack fall from his shoulders.

"We'll be staying here until morning. I suggest you get what sleep you can." He stepped over to a nearby tree and sat down to keep watch for a while.

The girl didn't move from where she'd stopped, just slouched down onto the ground, curled into a ball in the blanket and fell asleep. He shook his head then stared off into the night.

# CHAPTER 2

As THE SUN ROSE AND ITS LIGHT PERMEATED THE TREES, Torren stood up and stretched. His dream might have driven all thought of sleep from his head, but keeping guard through the rest of the night had let the time pass effectively. He'd long ago gotten used to sleeping little.

He reached for his pack and brought out some wrapped cheese and bread he'd bought from a farmer a couple of days earlier. This part of the empire was filled with farms and small towns, running almost to the border. The residents were usually willing to part with some of their stores for coin or labor. The prairie fields farther south produced most of the grain; wood, vegetables—mostly corn—as well as fruit were the contributions of this area.

Taking the food, he walked over to the blanket bundle on the ground and hunkered down next to it.

"It's time to wake up." He nudged her with the back of his hand then jerked back in shock as the blanket exploded and she sat up with a start. The girl darted her eyes in every direction, looking totally disoriented. Panic covered her face as she finally turned to look at him, and she appeared as if she might bolt.

"Forgot me already, have you?" he asked her with some sarcasm. "Run off, if you want, though I would have thought you'd rather have some breakfast." He tore off a piece of the hard bread and popped it into his mouth.

"You...You're the one who helped me?" She eyed him warily, as if afraid to believe this might be so.

He studied her, half-amused and half-annoyed, thinking surely he didn't look *that* bad. There were a number of women who thought him quite handsome.

"Do you want food or not?"

Slowly, as if afraid of committing herself, the girl nodded. He tore a chunk off the bread and part of the cheese
and held them out. After a moment, she took them, making sure she didn't touch him. She got up and, dragging
the blanket with her, shuffled several feet away before sitting back down to eat.

Torren ate his own meal, surveying his impromptu
company fully for the first time. She was young, so much
was obvious—no more than fifteen summers, was his
guess. Her hair was long, tied in a disheveled braid, its
sandy color much darker than his white-blond. Her face
was narrow, her mouth and lips small. She possessed
long, gangly arms and legs.

Her skirt was made of homespun and went down to her
ankles; but the cotton shirt was of better quality, with
sleeves that reached to her elbows. She also wore a vest of
dark brown with red flowers stitched around the border. A
blue clip at the end of her braid caught the light and
looked expensive.

Though a little better dressed than he would have expected, she still looked like a farmer's daughter. Overall,
she was unassuming and average-looking, her large sky-
blue eyes the only feature about her that stood out at all.

Nothing he saw explained why men would have chased
her into the night. Not that it mattered.

"Could I...Could I have a little more?" Her fear and
hesitation were quite clear.

He tore another piece of bread for her. "Thirsty?"

She nodded as she gingerly came forward to reach for
the offered bread. She took it and scooted back as he rose
to his feet. He felt her staring after him.

Torren took a deep drink then walked over to hand her
the waterskin. She took it eagerly. He stepped back,
watching her drink, wondering what he was going to do
with her.

"So, why were those men after you?"

The girl choked at the question, her gaze darting
around as if the mere mention of her pursuers would
bring them.

"Well?" He tried not to sound impatient but was having a hard time of it.

The girl set the waterskin down and stared at her lap. "I–I don't know." Her whole body tensed. "I was sleeping and my...my aunt, she woke me up and...and told me to dress. I asked her why, but she wouldn't tell me, she just told me to hurry."

Now that she'd started talking the words came out faster and faster.

"When I was done, I started toward the door, but she stopped me. She...She told me to go out the window."

Her eyes filled with tears. Torren suddenly felt uncomfortable.

"She pushed me toward it, telling me she loved me, telling me to hurry. She was whispering. She sounded afraid. It scared me, so I did as she said. When I had climbed out the window, she told me to run."

He frowned, not liking where this story was going. He told himself again this had nothing to do with him.

"I didn't run," the girl said, sounding utterly miserable. "I tried to argue with her. I knew something wasn't right, and I just couldn't go. That's when the door to my room slammed open, and my aunt turned around and attacked the stranger there." She took a tattered breath. "He...He hit her. She fell. And then...then I...I ran and ran, until..."

She stared at her hands, her voice shrinking to nothing.

"What's your name?"

She glanced up at him, looking surprised. "Larana."

Torren nodded. "And do you know where you are now, Larana?"

She stared at him for a long moment then slowly shook her head.

"All right, then," he said, folding his arms across his chest. "I'm heading north, in the direction of Caeldanage, and I'm willing to have you along until we either run across your home or come across a farm or town where we can find someone willing to take you."

Larana just stared at him, saying nothing.

"Of course, if you prefer, you can go wherever you want on your own."

She looked away, shaking her head vigorously.

He nodded. "By the way, my name is Torren."

Though she flinched as he came close, he paid no attention to her reaction and retrieved the waterskin. He went back to his pack. "If you're up to it, we should get going."

Larana nodded quickly and rose to her feet. After dusting herself of leaves and dirt, she grabbed the blanket she'd slept in the night before and briskly snapped it in the air twice before folding it neatly and then meekly bringing it over to him. "I'm ready."

He took the blanket without comment and wrapped it into a roll with the other, attaching them to the bottom of his pack. He glanced up past the trees, getting his bearings from the rising sun, and set off north.

He didn't lead them back to the road but stayed in the lightly forested area. The going was harder this way; but Larana didn't complain, though it was obvious at times she was hard-pressed to keep up.

When he called for a stop hours later she dropped to the ground in relief.

"Stay here."

"Where...Where are you going?" Larana straightened up, fear flooding her face as if she thought he meant to leave her.

He gave her a quizzical and slightly irritated look. "I'm going to lay a false trail. I'll be back soon."

She sat looking alone and forlorn as he left to take care of business. He hoped this wasn't an indication of a long and nerve-wracking trip.

He set about erasing as many signs of their passing as possible. Backtracking, he set off in a different direction, leaving clues that could be followed but not making them too obvious lest their pursuers realize what he'd done. As soon as he reached an area where a trail would be hard to find, he went back a different way, being as careful as he could not to leave any trace.

When he returned to where he'd left the girl, he found her pacing, scanning the area around her intently. As soon as she spotted him, her face lit up with relief.

"You're back!"

Torren scowled—he'd told her he'd return. He retrieved the waterskin from his pack and took a long swallow. As an afterthought, he offered it to her. In her eagerness to get it, she almost tripped over herself. His scowl deepened, but he said nothing as he handed it over.

Larana drank the water gratefully, her cheeks touched with red. "Thank you."

He shrugged and took back the skin. "Let's go."

After a short while, the leaf-strewn floor gave way to a small path intersecting their current direction. Torren stopped and glanced both ways then prepared to cross it.

"Wait!" Larana jumped forward and grabbed his sleeve. She immediately let go as he turned to glare at her.

"What is it?" he demanded.

His annoyance grew as the girl hesitated, staring up and down the trail as if looking for the right words.

"I–I think I know this path. It's a shortcut."

He waited for her to elaborate, but she didn't.

"To where?"

He watched as she bit her lower lip and glanced up and down the trail again, looking unsure.

"It's a shortcut to the stream," she said finally. "It's where we get our water." She pointed to the left side of the trail. "My home is this way."

Torren glanced down the way she pointed. "Are you sure?"

She bit her lip again. "N–No."

He studied the path. Though he suspected the men last night were even now trying to pick up her trail, there was a chance one might have stayed behind, waiting for her at her home to make sure she didn't return. Then again, it was almost as likely he hadn't. If her family was still there, though, he could leave the girl with them, freeing him to go on his way. Whatever problems her people were having with these men they could sort out themselves.

"All right. We'll follow it for a short while and see if it grows more familiar."

Larana nodded in thanks then took off to lead the way. He followed at a more sedate pace, shaking his head.

They'd not gone far before the girl turned around, a

bright smile on her face. "This is it! I'm sure of it now."

She ran, showing more energy than she had so far. As she moved farther and farther ahead, Torren slowed. A strange smell tainted the pervading scent of growing vegetation. Was that smoke? And what about the other, more subdued odor mixed in with it?

"Larana!"

He sprinted up the path, a sense of dread rising inside him.

After a long bend in the path, the trees opened into a clearing. He slowed as he spotted the girl at the end of the trail. She stood unmoving as he came closer, what she was staring at gradually coming into his field of view. The smells that had first alerted him something wasn't right grew stronger.

In the middle of the clearing, charred beams reached toward the sky, resembling broken, crippled fingers. Thin trails of smoke rose from the ruins.

Larana panted as she stared at the destruction.

"Is this...?" He left the question unfinished, knowing it could be nothing else.

The girl took a half-step forward, seemingly unaware he was even there.

"Aunt Ban? Uncle Zed?" Her call reverberated around the clearing, but she received no answer. "Aunt Ban! Uncle Zed!" she called out again, her voice fraying at the edges. "It's me, Larana. I've come back."

No answer disrupted the silence.

Torren felt his jaw clench, already knowing what she was yet unwilling to accept.

"They're...not here."

The girl turned on him, fire in her eyes. "They are! They wouldn't leave without me."

Turning from him, she ran into the clearing, heading toward a small shed on the far side—the only thing still standing. "Aunt Ban!"

He didn't follow her, instead approaching the burned-out shell of the house, sure he knew where the girl's relatives could be found. Following his nose, he moved carefully through the rubble until he found the source of the telltale odor mingling with the reek of smoke.

"Aunt Ban! Uncle Zed!" Larana's voice had grown shrill, filled with dawning panic.

He stepped out of the ruins. "I've found them."

She stopped where she was and turned to look at him, a hopeful smile on her face. He watched her look past him and said nothing as the smile slowly crumbled with inevitable understanding.

"No." She shook her head slowly from side to side. "*No.*"

Her expression despairing, she cut past him. He didn't try to stop her. He didn't watch as she stumbled into the rubble and shortly found the two burned and twisted bodies, which had, not long before, been her family.

"*No!*"

He turned at her cry, despite his original intentions, and saw her fall to her knees. He stared at her back as sobs racked her body. Without a word, he turned away from her pain and strode to the shed on the other side of the clearing.

Setting his pack outside, Torren searched inside and came out carrying a shovel. Not once glancing in the girl's direction, he proceeded to dig a hole not far from the side of the small building. Perhaps he could do for her what he'd not been able to do for himself.

Sometime later, he wiped his sweaty brow. Climbing out of the hole, he set the shovel aside and reentered the shed to retrieve several large pieces of sackcloth he'd noticed.

Larana still sat where he'd last seen her, her eyes red and swollen, soot covering her clothes and face, dark tracks showing the path of her tears.

"I've dug a grave for them," he told her.

She slowly turned her head toward him, her expression slack, her eyes glazed. It was hard to look at.

"If you'll move back, I'll wrap them up in this."

The sun was high in the sky, shining down on the manmade clearing. The stench from the bodies was growing stronger. Her face vacant, Larana blinked several times then crawled to her feet to get out of his way.

What debris there'd been over the blackened bodies was gone, though pieces of the corpses had come away with them. Suddenly, not wanting her exposed to this any more than necessary, Torren quickly laid a cloth over each one. His mouth a thin line on his face, he knelt, respectfully tucked the cloth around the body of what he presumed to be Larana's aunt and lifted it in his arms.

The stench of the charred flesh multiplied as the body shifted. Momentarily, he closed his eyes, unwanted images flashing through his mind of another time. When he opened them again, his vision was clear but his expression grim. At least these two would have the benefit of a proper burial.

Larana followed him automatically to the grave. She knelt in the grass, staring into the hole as he set the wrapped body inside it. Glancing once in her direction, he left her there as he went to retrieve her uncle.

After he'd settled the second body, he took a deep breath and spoke. "What gods did they believe in?"

She only stared at the grave.

He waited to see if she'd respond at all, but she said nothing. Sighing, he bent down long enough to take a handful of dirt and gently sprinkle it over the bodies. "May the First Mother take you to Her bosom and care for you."

He picked up the shovel and started filling in the hole. Larana said nothing as he worked, but fresh tears streaked her soot-covered face.

Once he was done, he took a deep drink from the waterskin then took a small piece of sackcloth from his pack. After dampening it, he used it to mop his face.

"How far is the stream down this path of yours?" he asked.

She stared at the mounded earth as if she could yet see the bodies lying within. She said nothing.

He shook his head and turned away. Taking a spare set of clothes from his pack and a pail from the shed, he headed across the clearing without another word.

Following the path, he soon came across a respectable stream. Setting the pail and his clothes to the side, he stripped and crouched in the cool water. Small fish nibbled his toes, but he paid no attention. As he washed his body and his dirtied clothes, all he could see was the soot-

17

covered gangly girl staring at her relatives' grave.

When he returned, Larana was exactly as he'd left her. Frowning and pushing back his damp hair, he studied her from the corner of his eye as he set the full pail he'd brought inside the shed. He came back out to loom over her, his expression blank.

"We'll need to leave soon," he said darkly. "We've already been here longer than is prudent." No reaction. "I've brought some water so you can wash yourself."

She gave no indication she'd heard what he said.

He reached down and grabbed her arm, yanking her roughly to her feet.

"We don't have time for this! They're dead. Deal with it." His voice was thick. "You've had time to mourn. That time's now over. Go clean up."

Her eyes widening with the pain in her arm, she stared at him without comprehension. Torren hauled her away from the grave and pushed her into the shed. He grabbed a piece of sackcloth, and after dunking it into the pail, thrust it into her hand.

"Clean up."

When she still did nothing, the water dripping from the cloth in her hand to the ground, he took her hand and raised it and the cloth to her face. She gasped as the wetness touched her skin, her eyes focusing for the first time.

"Clean up." He kept his eyes locked on hers, moving the cloth across her forehead. "Clean up."

She pulled her hand free, looking at the wet cloth held in it. "Yes."

She blinked several times, as if becoming aware by inches of her surroundings.

"I'll wait for you outside." He felt strangely relieved to see life coming back into her face.

Larana answered with a single nod, bringing the cloth back up to her cheek. He nodded back and exited the shed to give her some privacy.

He waited for her in the shade of a large oak by the shed, studying the land and wondering why so many farms were built the same. From the remains of the house, he knew it'd had no more than three rooms. It would have had a thatched roof, whitewashed sod walls and a central hearth for preparing meals and heating the house in the winter months.

A small garden in the back would have been for common vegetables, the actual crop fields farther off. A chicken coop would have been built against one side of the house; and perhaps they'd owned a few goats or a mule, though there was no sign of either now.

How similar it was to the place where he'd spent the latter part of his youth, a place that had been both a prison and a haven to him.

Shaking himself out of the strange, misplaced mood, he pushed away from the tree as he spotted Larana exiting the shed. Her face and arms were clean again, her hair damp and in place. Though she'd obviously also tried to clean the worst of the stains off her clothes, aside from wetting and smearing the soot, they didn't look much better.

She approached him rather meekly. "I'm done."

He nodded and studied the sky. "We still have a few hours of daylight left, so we should cover as much ground as possible before it gets dark."

She followed as he retrieved his pack. Though she appeared more like normal, he noticed the dark circles growing beneath her eyes.

"Do you have any other relatives near here?" he asked her.

She looked away, her eyes turning sad. "No. I have no other family." Her gaze strayed to the mound where her aunt and uncle were buried. "I—I'm a foundling. Aunt Ban and Uncle Zed found me on the road."

Torren felt his right eye twitch. This story was starting to sound just a little too familiar for comfort. "I take it they had no relatives, either?"

Larana shook her head no.

"I see." He felt the odd mood overtaking him again. "Let's go, then."

He hefted his pack higher on his shoulders and set off the way they'd come. Once they reached the point where they'd originally found the path, he didn't leave it but instead followed it to the stream. Once there, he took the time to refill their waterskin.

"Do you know if this connects to a river, or a road?"

Larana nodded quickly. "Yes. There's a road that runs east to west, some ways down. I wasn't allowed to go that far, but I did it once."

She looked guilty at the admission.

He had traveled through this area a number of times over the years and thought he had a pretty good idea where the stream would take them. "Come on."

He stepped into the water and followed its course upstream. It reached about halfway up his boots. Larana hesitated long enough to remove her slippers then waded in after him.

Though the afternoon was warm, the girl was shivering by the time he called for a short break. Her teeth almost chattering, she slipped on a rock while stepping out and fell to her knees, getting her skirt and legs wet as well as the shoes she'd carried. He frowned at her bumbling even as she looked up at him, her cheeks coloring.

After a moment, he went over and offered his hand to help her up. As they touched, he felt a tickling in the back of his head, and something akin to gratitude.

"I'm very clumsy. Sorry for the trouble."

He let go of her hand as soon as she was on her feet, shaking his head at the strange feeling. "I think we'll be able to reach the road before nightfall."

Larana nodded, trying her best not to look cold. The circles he'd noticed under her eyes earlier were noticeably darker.

He made a decision, and the slight scowl that was his normal expression disappeared. Though he wasn't aware of it, it erased years from his face.

"We haven't eaten since this morning. Why don't we stay here a bit longer and eat something to hold us over until we make camp?"

The girl nodded eagerly. "Yes, please."

He rummaged through his pack and pulled out a hunk of meat wrapped in waxed cloth. With his boot knife, he cut portions for both of them. She wolfed hers down after the first tentative bite. He was thinking of giving her more when she enthusiastically licked her fingers but hesitated as she abruptly stopped and tears formed in her eyes.

He knew loss was never easy, but it was best to just deal with it and then forget.

"I'm sorry about your aunt and uncle, but you need to put their passing behind you. There was nothing you could have done. Nothing will bring them back no matter how much you want it. For your own sake, just forget about them."

Larana turned to look at him, her face filled with shock. "How can you...?"

He stood up and slung his pack over his shoulder. "We'd better get moving."

In less than an hour, they found the place where the stream crossed the road. Thick planks had been set to make a small bridge. He climbed up, staring long and hard in both directions as Larana moved to join him, her still-damp shoes making squishing noises.

"Let's keep going just a little longer," he said after a minute. "Then we'll get off the road and set up camp." He eyed her; she nodded and said nothing.

They hadn't gone far before he turned off to the side. He penetrated the tall grass and brush just enough to get them out of sight. "This should do."

The girl sank down by a tree with a sigh and removed her shoes so she could rub her tired feet. He chose another tree nearby and removed his pack before sitting. He un-hitched the blankets and tossed one to her. He then removed the remainder of the meat as well as more hard bread and cheese, dividing most of it between them.

As they ate, the sun disappeared from overhead and everything plunged into deep shadows before being swallowed by darkness. He was caught off-guard as Larana, a mere lump of deeper shadow across from him, whispered, "Have you...Have you lost a loved one, too?"

He said nothing, not liking the question. There were things he didn't enjoy thinking about, let alone speaking of to a stranger. He grabbed his blanket and spread it out on the ground.

"You'd better get some sleep. We'll be starting out early in the morning." He lay down and turned his back to her, hopefully cutting off any further conversation. He stared into the darkness, listening to her settle in before eventually drifting off to sleep.

# CHAPTER 3

A HEAVY WEIGHT CRASHED FROM ABOVE, PINNING HIM AND *the others down. Dark-clad bodies descended on them from the closing gloom. Grinning bloodlust; answering fear. His father cut in front of him, blocking his view—hastily trying to push him back. His angry expression changed to one of abrupt pain. His father falling on top of him, forcing him down, warm liquid splashing on his face and arms.*

*Panic, madness. Screams from the others yet no way to escape. Pinned, not able to breathe. His fellows dying, others wounded. The dark men crippling them as they laughed at their predicament.* Why are they doing this?

*Suddenly, freedom is his; but before he can flee, they close in, pushing him this way and that. The hands—the hands reach for him, tearing at his clothes, at his body, drowning him with pain...*

Torren sat up, his breath coming in harsh gasps. Fear chilled him, the echoes of past pain flooding him. Slowly, very slowly, the true night congealed before him, reality reasserting itself. The dream dissipated into the past where it belonged.

With shaking hands, he pushed his clammy hair away from his face. It had come again. Why? It made no sense. It had been almost a year since the last episode, and now he'd had the dream two nights in a row. Would he never be rid of it?

He twisted where he sat, his sweat-soaked shirt clinging to him. Angrily, he shoved his questions aside and pulled it off, feeling in his pack for another. He didn't put the new one on right away, though, letting the night air cool him. When he felt calmer, more like himself, he slipped his arms into the sleeves. He was about to bring it up over his head when a soft touch caressed his back.

Goose bumps rushed up his spine, a strange tingling sensation suffusing his body. A queer combination of feelings rushed through him: worry, curiosity, sadness. For a moment, it was as if his body were paralyzed while his confused mind ran in frenzy through a dozen scenarios of bandits or creatures running across him in the night.

Then, he was free, the touch leaving him as unexpectedly as it had come. A strangled sob came from behind him.

Torren whipped around, his hand slipping out of his shirt and automatically reaching for the sword he'd left sheathed beside him.

"Those scars..." Larana's sorrow-filled voice was barely audible, yet it froze him as if he were in the grip of whatever had just happened again. He could barely see her outline in the darkness, her words coming as if from a disembodied voice.

He shook his head, struggling to free himself of his paralysis as he tried to make sense of what was happening. "They're nothing."

"That's not true!" She leaned forward, her voice filled with grief. "Pain...there was so much pain." She hid her face in her hands and wept as if his anguish were her own.

Torren stared, not knowing what to make of it. What kind of girl was this? How did she know these things?

"What did you do to me?" The question came out as a harsh accusation.

She didn't answer, weeping quietly before him. He reached for her arm, making sure not to touch her exposed skin.

"Answer me!"

She looked up; and though he could not see clearly, Torren felt her gaze cutting through him. He let go of her. Confused and angry, he moved back and half-turned away from her.

"What did you do?"

After several long moments, Larana finally gave him an answer.

"I–I'm not sure. It's just something that happens some-times. I didn't mean…I didn't mean…" Her hand shook as she reached as if to touch him.

"I suggest you don't do it again," he said gruffly. He moved even farther away from her and slipped his shirt on. His mind in turmoil, he lay down with his back to her once more, willing her to leave him alone.

She scooted away, sobbing softly. He wasn't sure if it was because of his anger, the loss of her family or his past pain. Why did he even care? He lay awake until the sounds of her weeping finally faded away.

# CHAPTER 4

TORREN TURNED IN HIS BLANKET THEN OPENED HIS EYES to the dawn. He yawned, feeling sleepy, until his attention drifted across the way and he spotted the girl. She was already awake, sitting up and studying him.

As soon as their eyes met, she looked away, her cheeks coloring. He sat up and turned from her with a scowl. All that had transpired the night before came flooding back; he felt no better about it now. The best thing he could do was to get rid of this strange girl as soon as possible.

He stood, grabbing the previous night's discarded shirt in the process. He peeked out of the corner of his eye at the girl and found she hadn't moved. Her sandy hair was back in place, and her blanket was folded beside her. How long had she been awake? How long had she been staring at him?

A feeling of unease crawled through him. He was encumbered with too many mysteries and not enough answers. Maybe it was time to change the situation.

"What did you do to me last night?"

He felt a slight sense of satisfaction when she stiffened at the question. She looked at the forest floor, at the trees, the sky—anywhere but him.

"I didn't do anything to you," she said hesitantly. "It's not something I purposely do. It...It just happens."

Torren hung up his shirt and rolled up his blanket. "What 'just happens?'"

Larana stared at her hands in her lap. "I feel things. Every now and then, I see things. But it happens when it wants to, not when I say," she added quickly. She glanced over at him, trying to gauge his reaction.

His uneasiness grew, though he couldn't say why.

"I see. And what did your uncle and aunt make of this power of yours?"

Larana looked away again, her eyes fogging a little at the mention of her dead relatives.

"They asked me not to use it if I could help it." She fidgeted. "But I can't always do that."

He took a half-step toward her. "You say you sometimes see things. Did you see things last night?"

His throat was tight.

She looked up at him as if sensing his discomfort. "No. All I got was the fear, the horror of what was happening. The pain." Her voice grew small, her eyes darkening with sadness. "Why would someone hurt you like that? You were just a boy."

Torren's hands bunched to fists at his side, his whole body tense. How did she know this? Was it just a guess? Who was she? Did she see more than she was telling? It took an act of will to not think about it.

"We should be on our way. We can eat while we travel."

He turned his back on her and stiffly picked up his pack. He almost jumped when he turned around and found her close, holding out her blanket to him. He took it, not looking at her. Yes, the sooner he got rid of her the better.

Once ready, he led the way back out onto the road and headed east. He took out some food and passed Larana back her share, barely looking at her. She tried to smile in thanks, but he pretended not to notice. He'd already involved himself with her problems more than he liked.

About midday, the road they were on connected to a much larger one paved with stone. It was a Grand Highway, legacy of the Emperor Solarious. The highways had been the emperor's way of unifying the empire while at the same time guaranteeing his name would be spoken for generations to come.

Stepping onto the highway, Torren glanced north and south and spotted what might be a caravan coming their way in the distance. The main merchant routes all lay along the highways; they were the most direct means to the majority of the larger cities and, with the garrisons and outposts set along the way, one of the safest routes as well. Which was exactly the reason he tried to avoid them. Only high-priced inns and vendors could be found along the highways. Torren could eat for days on what one of the inns charged for a single meal. How the pilgrims, who used the roads on their way to the multitude of holy places of the Goddess and her children, could afford them he had no idea.

Unfortunately, due to his present circumstances, using the highway would be his best course. With a small sigh, he turned and headed north.

A few hours after midday, the caravan caught up to them. They moved off the road to let the merchant and his wagons pass, using their arrival as an excuse to take a rest. He studied Larana as she inched toward the highway's edge, intent on the covered wagons laden with southern spices and silks. She waved eagerly at the drivers and the men walking beside them. Some of them waved back.

He also studied the group but for different reasons as he noticed the high number of guards they'd brought along. It looked as if he wasn't the only one aware of the rumors about the massing army to the north. War could be good for business, but it also brought the vultures to the fore.

Once he left Larana somewhere, he might consider selling his services part of the way to the border to one of the caravans. The pay wasn't bad, and the work normally easy. It would do until he could hire himself out to the army, if there was one, or, better yet, sell his services to a private party seeking a little protection as they fled the potential conflict. Even the empire employed mercenaries for certain jobs now and again; and with the mountains and the narrow passes, he was sure the struggle, if one was truly in the hatching, would be long and taxing.

Once the caravan was past, he got up to start on their way again.

"Is it exciting to travel to other places?" Larana asked, her eyes bright. "Are the southern kingdoms as strange as they say?"

He looked into her expectant face and shrugged.

"That would depend on your definition of strange," he told her. "Most places are alike. The climate might be different, the style of dress, the food, but inside, people are the same no matter where you go." He set off, leaving her to catch up.

"I'm sure that's true, but they still *feel* different, don't they?"

He sighed. "I take it you've never traveled?"

She rushed up to his side. "No, only to local festivals. The farm is all I've ever known. My aunt and uncle weren't much into travel. Said it was a waste of money, and the fields wouldn't tend themselves while we were gone. But I've always thought it would be exciting. You've traveled a lot, right?"

Torren nodded. He'd been traveling since he was sixteen. He'd seen more than his share of what the world had to offer. "Some."

"Please, won't you tell me a little about other places? Just for a while?" she begged.

From the look on her face, he doubted he could put her off very easily. At least it would help pass the time.

"The Southern States of the Empire were some of the last additions before the Time of Peace began over a hundred years ago. The weather there is a lot more humid and hot than here. The cities bordering the sea are cut into huge cliffs and overlook the water. The people dress in sheets, wrapping themselves up from head to toe."

"That sounds very hot and uncomfortable."

"Their fabric is a lot lighter than what you're used to. They use thread spun by insects instead of wool." He could tell by her expression she'd never heard of such a thing. "The Eastern States are also very humid, covered with jungles and full of strange creatures. The people there are deeply into the mysteries, and feel that every breeze, every animal, can be used to interpret the wishes of the gods.

"To the west, the land is harsh and difficult to live on. The men there are hard and the women harder. The empire's mines are there, and many go hoping to find their fortunes, though often what they meet are bandits or worse."

"Where are we?" she asked.

The question caught Torren off-guard. "Merris, one of the Northern States. As long as we don't invade and take over Galt, that is."

"I see."

As they walked, the trees gave way to tilled fields. Not much farther on, the low wall of a small town could be seen. An inn stood just inside the wall, a paddock big enough to handle the wagons and horses of caravans in the back. A small imperial garrison was across from the inn, their facilities almost as large.

Beyond the inn and the garrison were a general store, a blacksmith shop and a farmers' market. A number of houses dotted the area, most sited away from the road.

"Will we be staying at the inn?" Larana asked hopefully.

Torren glanced into the filled yard. "There won't be room. The caravan beat us here."

He didn't tell her merchants meant wares, selling and questions. He still had no idea who had attacked her family or why, and associating with a bunch of nosy men wasn't his idea of keeping a low profile.

He came to a sudden stop, Larana almost bumping into him. Wait a minute, why wasn't he thinking? His gaze riveted on the garrison; the place was exactly what he'd been hoping for. He would drop the girl off here, let her tell them what had happened; and they would become responsible for her instead of him. If he asked her to, he was sure she'd keep his involvement out of the story. He would be free to go on his way.

As the idea grew in appeal an unwanted question sprang up. Once the report was investigated, and they found out she possessed no living relatives, what would they do with her? She had no money. She was too young to rebuild her home or even care for the farm. He glanced back at her. What was she—fifteen? Certainly, no more than sixteen. He doubted she had much education or experience, nothing she could use to earn an honest living.

The gangly girl smiled shyly at him, not understanding

what they were doing standing in the middle of the road but waiting for his lead.

He looked away from her and stared at the garrison again. It wasn't his problem. He'd done more than could have been expected of him by bringing her this far. But...

"Torren?"

"Come on." He faced the road and went on, leaving the garrison behind.

On the right, a few stalls in the farmers' market were in use, but most were empty. He knew during the harvest season and the ensuing festivals farmers from all around would come here to celebrate and sell their produce and wares. If one had been going on now, he might have possibly been able to find a family willing to take Larana, with some monetary incentive, of course. The fact she knew her way around a farm would have been a plus, or at least he would have said as much. He actually possessed no idea what she could or couldn't do.

The offerings at the manned stalls weren't much—a few fruits still in season and some vegetables—but he looked through them as if interested. The bored farmer brightened.

"Care for some luscious fruits and vegetables, sir?"

"Perhaps."

Larana stepped up beside him and started going through the stock with a practiced eye.

"And there might be something else as well."

The farmer stared at him, suddenly intrigued. "You don't say."

"How much for these?"

He sent her an annoyed scowl as the farmer's attention was diverted.

"Why, those beauties are two for a five-piece," he said. "You won't find any better."

Larana stared critically at the tomatoes in her hand then glared defiantly back at the vendor. "A five-piece for two? No self-respecting person would pay that for these. Look, they're still a little green, and they're so small. Three for a copper is more like it."

The farmer cringed. Torren stared at the girl as if he'd never seen her before.

"You wound me, miss," the farmer said. "Surely, you can appreciate the expense for quality. Three for a five-piece."

She snorted. "You're a thief and a scoundrel!"

Torren took a step back as the girl put her hands on her hips and stared the old farmer down. What was this? Where was the shy waif he'd rescued a few days before?

"A pup would know these aren't worth more than a half-copper each," she asserted.

Instead of getting angry, as Torren was sure he would, the old farmer cackled for a moment, his eyes bright. He then gave back as good as he'd been given. In a flurry of words bouncing back and forth, the two of them got down to a serious session of haggling.

After a few more minutes, Larana turned and gave a startled Torren a triumphant smile as they settled the price to three for two coppers. Shaking his head, he reached into the side of his pack for the necessary coinage.

"Thank you, sir," the farmer said as he took the money. "Might there be anything else you'd be wanting?"

Torren nodded, taking a moment to remember why it was he'd really approached the vendor in the first place. "Yes, actually, there is something else. Might you know if one of the locals would be willing to lend us some space in a loft or a stall for the night? The inn appears to be full at the moment."

"Well, I might be able to help you there, sir." A twinkle returned to the old man's eyes. "I might just have room for you, I might. But I couldn't let the precious space go for less than a gold."

His gaze slid from Torren to Larana, his expression expectant. He wasn't disappointed.

"Are you mad? Even during a festival no one would ever think of charging such outrageous prices!"

Torren watched the ensuing battle with mounting amusement. He'd not realized haggling could be so lively, or its participants so happy. He, personally, didn't have the patience for it. These two gave the impression they lived for nothing else.

After several minutes, the farmer and Larana finally settled on a price. Smiling, the older man gave a nod to

Torren. "I guess since the two of you will be staying at my place, it'd only be hospitable to invite you to dinner."

"That would be very welcome," Torren replied as he fished out the agreed-on fee.

Larana looked smug.

"Let me get my wagon, and after you help me close up, we'll go." The old man winked at Larana then left them to watch over his goods.

"Where did you learn to do that?" Torren asked once the farmer went out of sight.

Larana looked suddenly embarrassed. "From my uncle, at the spring and harvest festivals. It's a tradition. Competition is fierce, and it's fun. I did most of it last time." Her eyes lost a little of their light. "I'd been looking forward to showing him how much better I was going to do this year."

"You're very good."

The girl nodded at the compliment but said nothing else.

Once the farmer returned, they helped him load his wares into a small mule-drawn cart and walked on either side of it as he took them to his home. The farmer, who'd informed them his name was Gimmel, took a narrow dirt road at the back of the market area and headed for one of the houses Torren and Larana had spied from the highway. Gimmel's was a whitewashed two-story with a large barn sitting beside a neat spread of fields.

"If you wouldn't mind taking Cully here on to the barn, I'll go tell the missus we have guests for dinner." He gave them a big smile.

"Leave it to us." Larana gave him a smile back, taking the mule's bridle. Tugging on it to get him going, she led the animal in the direction of the barn. Torren followed.

By the time he entered, Larana had already figured out how to unhitch the cart. She scratched the mule affectionately behind the ear, whispering softly to him as if sharing a secret. The animal brayed and sedately clumped over to his stall without giving her any trouble. After taking off his bit and bridle, she found a brush and, still talking to Cully, brushed him down.

Torren watched her from the corner of his eye as he set

his pack down in an empty stall then unloaded the cart. In his experience, mules were stubborn, cantankerous creatures, yet this one seemed to be no trouble at all. Looked as if she might just have a few more skills than he'd given her credit for.

The two of them were just about finished with their chosen tasks when Gimmel entered the barn with a young man in tow.

"I've told the missus we have company, and after I showed her your coins, she was more than happy to hear you'd be joining us for supper." He gave them both a grin. "This here is my son, Acer."

"Pleased to meet you." Acer bowed his head in their direction, his square face as open as his father's.

"Why, you've already taken care of everything," Gimmel exclaimed with some surprise. "Cully gave you no trouble?"

Larana shook her head. "No trouble. He's great!"

Torren saw startled glances pass between father and son. After a moment, Gimmel shrugged and went on.

"Well, I may not have gotten as suckered as I thought, then." He openly smiled at them both. "Come on inside, and I'll introduce you to the others." He led the way over to the house.

The front door opened into a large common room. In the center was a broad table surrounded by wide chairs. A long bench sat to one side, and a loom resided in the back. An open fireplace took up much of the left wall, and the small fire currently lit in it kept the room warm. Curios and knickknacks decorated the mantel, welcoming inspection.

In a deep chair beside the fireplace, a young woman was softly cooing to a sleepy baby. She looked up and smiled easily as they entered, and an older woman emerged from a doorway on the right carrying a large tray.

Larana suddenly turned shy and hung back. Torren stood impassively, as was his usual wont, leaving Gimmel to make the introductions.

"Macah, Ulla, these are our guests...ahem." The old farmer took off his hat as his face suddenly colored. He held it in his hands, crinkling it. "Heh, I guess I never did get as far as asking your names, did I?"

The older woman, Macah, set the tray on the table and raised a thick brow.

"Gimmel, I swear if it wasn't attached..." She left the rest unsaid. Ulla tried to hide a smile.

"I'm Style, and this is my cousin, Leila." Torren sent a warning look at Larana as she stared at him in astonishment.

Macah motioned them to come into the room. "Well, Style, Leila, it's good to have you. From what Gimmel tells me, Leila forced a good deal out of him." She gave them a conservative yet warm smile. "He does so miss the verbal contests outside of festival season. Please take a seat."

Torren took a chair at the table, Larana quickly sitting beside him.

"What happened to your clothes, dear?" Macah asked as she got a good look at what Larana was wearing.

"Uhm." She hesitated, glancing to Torren for guidance.

"She chased a rabbit into a field and fell into an old fire pit that hadn't been covered over very well," he said, as casually as he could manage. "She tried to clean it with some water, but..."

"Oh, that won't do." Macah stared at the clothes critically. "I'm sure I can get it out. That vest is too pretty to leave like this. If you don't have a spare set of clothes with you, I think Ulla and I can come up with something to hold you while we clean it up."

Larana blushed. "Thank you, that would be very nice."

Macah clucked the thanks away and started serving everyone. The portions of roast were stretched a little, but Macah had enough extra vegetables, bread and cheese to provide for the unexpected number of people at supper.

Gimmel and his son discussed the day's business while they ate, and the fact a new caravan had come to town.

"Sales might be good tomorrow, then," Macah commented. "You'd best get there early in case they wish to buy supplies before they move out."

"Sure, but I don't know if they'll be buying," Gimmel said. "This lot, just like the last, seems to have an awful lot of guards with 'em. The merchants have been pretty prickly of late, looking at everyone as if they'd all become murderers or something."

For the first time, Torren became interested in the discussions. "Might you know why?"

Gimmel flashed him a smile then leaned back to stare momentarily at the ceiling, as if gathering his thoughts. His wife rolled her eyes.

"Well, I reckon it has something to do with the rumors." He paused, glancing at Torren with bright eyes.

Torren took his cue and asked, as expected, "What rumors?"

"Ah, well," Gimmel said, warming up to the subject. "The way I hear it—the rumors, that is—there's some kind of build-up going on a ways past Caeldanage. Some say it's an army of some sort, though others say it's a big con to make people panic and drive up prices. All I know is an awful lot of people seem to be suddenly interested in the goings-on up there. Heck, even the garrison here has had extra men come up in case of trouble."

Torren leaned forward. "What do *you* think is going on?"

The older man's smile turned suddenly sly. "Well, now, that's a mighty interesting question."

"Gimmel, don't go telling these nice people that foolishness," Macah cut in. She stared at her husband, her look partially indignant and partially annoyed.

"Now, wife, you don't know it's any such thing! Isn't that so, son?" He turned to Acer for support.

Gimmel's son just threw up his hands and kept chewing on the large piece of cheese he'd hastily thrown into his mouth. Torren wondered how farfetched the old man's theory could be that they so obviously didn't want him repeating it.

"You all just take the fun out of life," Gimmel grumbled. He sat back and pulled out a small pipe from his breast pocket, pouting.

Macah sent her husband a satisfied look and started picking up the dishes. Not quite hiding a smile, Ulla handed the baby over to her husband and got up to help her mother-in-law.

"I'll make sure to have plenty for breakfast, so be sure to join us in the morning," Macah told them.

Torren stared at her, surprised. "That's very kind of you."

Most farmers he'd ever boarded with had rarely invited him to supper, let alone breakfast—not for free, anyway. He saw Macah send a small smile Larana's way as she thanked them as well, getting up to help with the dishes. He suddenly understood why the invitation had come.

He was left with his thoughts as the women departed. Gimmel was chewing on the end of his pipe rather than smoking it, and Acer was busy rocking the baby.

Gazing about him, he recalled many nights resembling this one—good food, quiet company, a warm fire. How long had it taken before these things became comforting rather than painful or strange? He shook his head, not wanting the thoughts, for they always brought such a varied mixture of emotions with them.

Larana returned after a short while wearing a shirt and skirt just a little too big on her. Macah was behind her.

"Dear, why don't you show these good people where the well is so they can wash up—and get me some water, while you're at it?"

Gimmel's face lit up a moment, but it went back to normal before he answered her with a half-put-out "Sure." Grabbing a lamp from beside the fireplace and lighting it with a taper, he handed it to Torren then grabbed the pails his wife handed him and headed for the door as if unhappy to have to do the chore.

When he stepped outside, however, he straightened considerably and stared off in the direction the sun had set. Now that it was just the three of them alone, Torren expected the old man to pick up the conversation where it had left off. Instead, Gimmel kept silent as he took them to the back of the house and pulled up some water to fill both containers.

"This one you can take back to the barn with you."

He escorted them, dropping his pail off on the way. They entered the barn, and Larana tried not to trip over her long skirt as she hurried off to the stall they'd be spending the night in. She quickly tried to smooth out the hay and set out the blankets while they still had the light from Gimmel's lamp.

As she worked, Gimmel sidled up to Torren, his eyes bright. "It's my belief, though, as you can tell, my family doesn't agree, the goings-on up north have to do with the Flyers."

Torren felt himself stiffen. "Flyers?"

"Yeah, that's what I think." His voice grew even softer. "I also think it might have something to do with the Vassal of El." The old man watched for his reaction, nodding his head knowingly.

Torren stared at the old man, numb. The Vassal? After a moment, he shook himself out of his surprise. Could what Gimmel said be true? His family obviously didn't think so. The man offered no proof; it was foolish talk. Still, the uneasy feeling in his gut wouldn't go away. A quiet voice in his mind reminded him the dream had returned. Maybe it wasn't just coincidence.

Seemingly having gotten the reaction he was looking for, Gimmel went on. "I, for one, don't believe the stories told of the Flyers. Do you?"

"Not exactly." He had heard many different stories told about them. Eaters of lost children, bringers of bad weather. How they'd supposedly stolen the favors of the gods for themselves.

"And you shouldn't, either, though I'm in the minority around these parts." He lowered his voice. "I actually saw one once, at the capital, a long time ago. They didn't look evil to me. If anything, the ancient stories told there said they were once caregivers to all who had need of them. That if you lived life like them, you could actually become one of them."

Torren had heard those tales as well. While the first was close to the truth, it had been a long time ago. As to the second...

"Well, if you're all set, I'll be bidding you goodnight. Wouldn't want the missus to start wondering what I'm up to." He grinned shamelessly.

Larana looked up from where she'd finished setting out the bedding and waved. "Goodnight."

Torren walked over to the stall as Gimmel opened the barn door. He sat just as the barn was plunged into darkness. Cully brayed, objecting to the interruption of his well-earned sleep.

Taking off his boots and lying down, Torren tried not to think about what the old man had said, being troubled enough already. He almost jumped as Larana's quiet voice reached him in the darkness.

"Torren, what are Flyers?"

He scowled, for more than one reason, at the question. "They're the Chosen of El."

Surely, she knew this.

"El?" she asked.

Did her aunt and uncle teach her nothing? In all his travels, he'd never before met someone who'd not heard of the Flyers or El. "He's one of the sons of the Goddess. The Chosen, or Flyers, as they're normally called, are his people. Supposedly he blessed them long ago and took them as his own after they'd done him a great service back when the world was young."

He almost said more but made himself stop. These things weren't important.

"The Chosen of El." Larana's disembodied voice sent an unexplained shiver down his arms. "Do they eat children?"

He sighed. "No. They never have, and I doubt they ever will. It's an old tale told by people with no understanding of what they're talking about."

"And the other thing?"

"Macah was right. The old man was only talking foolishness." He cut her off before she could say anything else. "Go to sleep. We'll be getting an early start tomorrow."

He lay staring into the darkness until he finally heard her settle down. Only then did he allow himself to sleep, old stories whispering in the back of his mind.

# CHAPTER 5

**T**ORREN WOKE THE NEXT MORNING AS PREDAWN LIGHT peeked in through the open barn doors. He could hear a barrage of soft curses coming from the mule's stall.

He sat up slowly, brushing stray strands of hay from his hair, amazed to have slept the night through. Since the dream had plagued him for the past two nights, he'd assumed it would return for a third. Maybe whatever had triggered it was gone. He could only hope so.

The mule brayed in annoyance as Gimmel tried to coax it out of its stall. It tried to kick him. The nimble old man knew the mule's ways well and made sure not to be in range.

"You confounded pest!" He halfheartedly made as if to swat the mule then turned from the stall in disgust. He spotted Torren sitting up. "Ah, good morning," he said. "Sleep well?"

"Fine." For the first time, he realized Larana wasn't with him. A sliver of apprehension cut through him.

Gimmel beat him to the question before he could ask it. "The girl woke up when I first came in. She offered then headed off to help the wife. Quite handy, she is, unlike some." He glared back at the mule. "You don't think she'd be inclined to help me get this cantankerous creature hitched up after breakfast, do you?"

Torren stood, grabbing his blanket and shaking it free of hay. He noticed Larana had already rolled hers, attaching it to the pack. He was amazed he hadn't heard her. He normally didn't sleep so soundly.

"I'll ask her, if you want."

"I'd really appreciate that. Indeed, I would."

He slipped on his boots then went to the pail of water Gimmel had given them last night. He splashed his face and ran his wet fingers through his hair. Gimmel was out of sight in one of the other stalls, so he didn't bother to pretend to shave, instead just cleaning his teeth. He then took the pail and dumped the water outside.

Uncharacteristically, he stared up at the brightening sky and took a deep breath, reveling in the fresh morning air with the hint of baking bread curling through it. He caught sight of a lone falcon drifting high on a thermal. With a yearning he hadn't felt in some time, he watched it glide in lazy circles above him. After a minute or so, he forced himself to look away.

Straightening the collar on his loose shirt, he grabbed the empty pail and headed for the house. The clatter of dishes and laughter greeted him at the door. The baby squealed with pleasure, and he spotted Larana tickling it as she made funny faces. She had her own clothes back on, and they were clean.

He almost smiled at the scene until he realized what he'd been about to do and sobered. None of this mattered.

"Morning!" Larana's greeting was bright. He only returned it with a nod.

As soon as he took a seat, Macah seemed to appear out of nowhere and served him breakfast—hot bread, fresh butter, cheese, milk, even eggs. Ulla teased her husband as he walked in yawning and then served him. The atmosphere was pleasant, even sweet. It rankled.

He glanced over at his charge and saw she looked happy, not like the frightened girl he'd first come across. With a wide smile on her face and light in her eyes, it was easy to forget her gangliness, her strange power and the fact he was trying to rid himself of her. He looked away and ate, his food not tasting as pleasant as before.

Once Torren was done, Ulla took her happy baby back from Larana, cooing at it. Larana got up as he did and followed him back to the barn to get his possessions.

"Gimmel wondered if you'd mind getting the mule ready for him," he said in a noncommittal tone. "I told him I'd ask."

Larana rushed forward then turned around while still walking and almost tripped.

"Sure, Cully's sweet." She grinned and ran off to the mule's stall. By the time he retrieved his pack, she had already coaxed the mule out and got him headed toward the cart. It was almost as if the two of them had been working together for years.

"These are nice people." Larana glanced over at him as she got Cully to back up.

"Yes, they're very nice." He forced himself not to look at her happy, trusting face and went on. "I could ask them if they'd let you stay."

"No!" She looked as shocked by the vehemence of her denial as he was. She turned her face away. "Please don't."

"Why?" he asked her bluntly. "We agree they're nice people. They live like you're used to. They seem to like you."

He was hard-pressed to feel unmoved as she looked at him, her face marred with guilt and pain.

"Did I do something wrong?"

"No. You've done nothing. But your family is dead. You have no living relatives. And my kind of life is not for you."

"But..." Her eyes filled with tears, yet they didn't fall. "It would be wrong to stay here. Please don't make me. It's too close..."

She stared at him in supplication.

He frowned, disturbed by her choice of words. Might it have something to do with her strange talent? Yet it made no sense. The men who'd been after her would be long gone by now. She'd not gotten a good look at them, and he doubted they'd had a chance to get one of her. They'd have no reason whatsoever to continue looking for her. Still...

"All right. Have it your way."

"Th–Thank you." Larana wiped at her eyes. Her relief was almost palpable.

The tension he had been trying to ignore in his shoulders suddenly dissipated. He didn't understand it and just told himself it was relief at having averted a possible ugly scene.

"I'll wait for you outside."

Gimmel showed up a few minutes later with his son in tow. He brought a wrapped bundle with him.

"Will the two of you be going soon?"

"Yes," Torren told him, "we still have a fair way to travel."

"Well, the wife figured as much and decided she wanted you to take this along." He gave Torren a grin and handed over the bundle. "It's just a little something to make the trip more pleasant."

He took the offered gift, nodding his thanks. The door to the barn squeaked open, and Larana came out with Cully, the hitched wagon trailing behind.

Gimmel shook his head, his grin growing wider.

"See, Acer, I told you she could do it." He gave Larana a friendly pat on the back. "My mule is sweet on you."

She scratched Cully behind the ear as she gave Gimmel a small smile. "He's a good mule."

Acer snorted.

"Even so, thank you for your help with him. It sure makes me feel as if I might have come out a little too ahead in our negotiations just for this."

Larana shook her head, turning serious. "Oh, no, I got a fair price."

She smiled again, but Torren couldn't help noticing how the smile didn't quite reach her eyes.

"Well, if the two of you ever come back this way, you make sure to come see us. Especially if it's during festival." He winked. "I think between the two of us we could rob most of the folks blind."

She grinned with actual pleasure this time, her cheeks flushing. "I'd like that."

"We'd best be going." Torren nodded to Gimmel and his son. "Thanks again for your hospitality."

He handed the wrapped bundle to Larana and headed toward the highway.

"Goodbye!" She waved at the two men, and the women now in the house's doorway, and ran after him.

As they reached the highway, it was obvious some of the caravan workers were already awake and hard at work. The neighing of horses could be heard as they were brushed and made ready to go. Several of the guards stared at them in distrust as Torren steered toward the general store.

"You'll need a change of clothes," he said as they stepped inside. "See if they have anything that might be appropriate."

The proprietor's face brightened at their entrance, a young boy with him looking half-asleep. Like Gimmel, it was obvious they planned to take advantage of the caravan if they could.

Larana instantly headed to the back of the store. He stayed by the doorway, taking stock of the goods on display, knowing this would probably cost him more than he wanted to spend. He wondered why he was bothering.

About ten minutes later, Larana came back with a simple dress. "Is this all right?"

He barely glanced at it and nodded, not looking forward to what would come next. As soon as she'd gotten his approval, however, she took the dress to the waiting proprietor and, before he could even open his mouth, stated clearly that she'd give him a copper for the dress.

The proprietor's brows rose at the ridiculous offer.

"I'm sorry, but it'll be two silver."

"Is that so?" she countered.

The proprietor's brows rose another notch. "Yes." The boy beside him looked more awake than before.

Instantly, Larana took up the challenge, pointing at the seam of the dress. The battle was joined.

Less than five minutes later, Torren paid a mere pittance for the purchase, amazed once more at the bargaining skills of this strange waif.

# CHAPTER 6

THEY WERE SILENT AS THEY LEFT THE BOUNDARY OF the town at a leisurely pace. After a time, trees crowded once more against the road, providing cool respite from the sun. He came to find Larana's continuing silence disturbing, yet he did nothing change it.

When they stopped to rest about midmorning, he watched her surreptitiously as she clutched the bundle given them by Gimmel's wife, staring back down the road the way they'd come. More than once, he considered asking her what she was thinking about, but didn't. It wasn't any of his business, after all.

When the sun glared down at them from high in the sky, Torren veered off the highway and settled in the shade of a large oak. Without being asked, Larana opened the bundle; and for the first time, they got to see what they'd been given. Nestled inside were a couple of loaves of fresh bread, a large chunk of goat cheese and several ripe tomatoes.

Larana laughed as she split the contents and gave him his share. "They really were nice people!"

He bit into one of the tomatoes, its mild juices tickling his tongue. "Don't you think you could have been happy there?"

The question left his lips before he could think about why he was asking it.

Her eyes went dark. She turned away, hiding her face from him, and suddenly, he regretted asking the question. When she answered, he had to work hard to hear her reply.

"I think I could have been very happy...but...I didn't want to take the chance I might bring them pain."

Was this paranoia, or something to do with her strange power?

"Besides," she added in an artificially cheery tone, "this way I get to travel and see things I've not seen before." She still wouldn't look at him. Torren said nothing, eating his food. She eventually reached for some of her own but still kept her face averted.

So, he was slightly taken aback when, later in the afternoon, she picked up her pace until she was walking beside him. He decided to say nothing about it, glad to see her face bright again.

Over the next three days, they fell into a companionable routine. They lit no fires at night by unspoken mutual consent and left the road whenever they spotted anyone coming their way, which wasn't often. The farther they went from where he found her, the more her spirit seemed to leave behind her sorrow. Occasionally, she'd dart from side to side on the road in surging spurts of energy as she chased after a bird or a squirrel or admired a beautiful butterfly or flower. More often than not, in her hurry, she would trip and, occasionally, fall. The first few times she picked herself up and went on, he shook his head or rolled his eyes. After a few more, an occasional grin tugged lightly at the edge of his mouth at her perseverance.

Larana quickly volunteered for anything that needed done, growing bold enough to take their blankets from his pack the moment he set it down and set them out for the night.

Their travels were quiet, almost seductively pleasant. If not for the one evening he was awakened by the sound of hooves striking stone as speeding horses rushed by on the highway and the occasional glimpse of the skyborne island far off in the distance, he might have thought the journey ideal.

On the third night, he found it hard to go to sleep after having glimpsed the island once more floating ahead of them. He stared straight up at the stars, wondering what it was doing there. It could in no way be related to

46

Gimmel's fantastic theory. Still, it was unusual to see one of them remain in the same position for more than a day, and this made him worry.

The islands had established paths and rarely deviated from them; the flying ships normally took care of any side tours the inhabitants needed to make.

With any luck, the island would be well on its way by the time he and Larana made it to Caeldanage.

On their fourth day, the trees thinned around them, giving way to large, open fields. Their view no longer obstructed, the large floating island hovered before them, a dark blot in the sky. Torren couldn't keep his eyes off it, a heavy feeling of foreboding filling him as it appeared it wasn't going to leave.

"Torren, what is that?"

Even before he glanced her way, he knew what Larana was asking about. He wasn't mistaken—she was pointing straight at what he'd been looking at all along.

"What do you think it is?"

She glanced sharply at him, obviously surprised he'd answered her question with a question. "I–I don't know. I've seen it before, or others like it, but never this close. Aunt Ban would only say they were the same as leaves in the river and float randomly from place to place—a wonder to be watched from far away. But this one isn't like those. Are they something more?"

He frowned. Had those who raised this girl been ignorant? Why hadn't they taught her what every other child in the region already knew, regardless of whether they believed any of it or not? He shook his head, not wanting to follow the strange track of thought.

"They're a gift from El. It's where most of His people live."

"The Chosen?" she asked, her eyes growing wide.

"Yes." He gazed once more at the far-off island. "The First Mother gave of herself to create the world. She then created the other gods, then the animals and, finally, man."

Larana turned to stare at him raptly. He watched her reaction from the corner of his eye. Had she heard none of this before?

"Having given so much of herself, the First Mother sat back to watch her handiwork grow and prosper. The other gods watched with her, until they grew impatient and came to the world." He was amazed at the ease with which the words came—it had been a long time since he'd thought about any of this. They were just tales, after all.

"The First Mother was still too tired from her labors to interfere and so didn't stop them.

"The gods played with the animals and the humans. Sometimes they were kind to them; at other times they were not. The gods were young, new, the same as everything around them. They didn't understand the cycle of life and death, for they, like the Mother, would never die.

"Time passed; the gods grew. They picked different areas they wished to hold domain over. Valem chose fear and death, Tani chose birth and hearth, Suw chose luck and mirth, Ran chose warriors and conflict, Yeska chose knowledge and magic, Talloon chose nature and miners and Zoole chose greed and money. All chose, all but El. He, like the others, possessed no real understanding of the humans but, unlike them, he wanted to rectify this problem, He wanted to know why these creatures, made by the Mother, who'd also made him, acted as they did."

He glanced at Larana and saw she was captured by the tale. At the moment, she looked younger than her years, the wonder on her face that of an infant. It plagued him, but he went on.

"El left the world and sought the Mother. He told her of his need to understand and the one way he'd come up with to reach this understanding." He paused a moment. How many times had he heard this part of the tale himself? "It is said the Mother was so proud of her son, so full of joy at his decision, that the night almost seemed like day, her smile was so bright."

Larana stared at the sky around her in awe, almost as if she could see it. He went on.

"Without hesitation, She granted her son's wish and sent him back to the world as a human, so he might live as one of them and thereby grow to understand them. El came to the world to find everything different from what he'd known as a god. He was alone, something he'd not experienced before—and felt lonely. The sky grew dark, and he found he could not see. For the first time, he experienced hunger and felt cold.

"By this time, a number of the other gods had learned the source of the Mother's pleasure and what El had done. All watched what El was going through—all but Valem, who felt cheated in that it hadn't been he who so pleased the Mother. Besides, why was it so important to understand humans, anyway?

"Thinking El a fool, Valem used the power he'd claimed on his brother. In the darkness, he gave El the gift he'd given man, the gift of fear."

She gasped. Torren stared at the floating island, amusement and bitterness growing within him at the same time. He could clearly remember his own reaction the first time he'd heard the tale. It felt strange to now be the one telling it.

"Suffused with this heretofore unknown emotion, El ran in panic through the dark, bereft of all his power, not even having speech with which to call his Mother. He ran and ran, smacking into trees, cut by thorns, chased by both real and imagined dangers, until he fell, exhausted and beaten.

"When he eventually awoke, he was in a bed, warm and safe. A young boy had found him, and those of his village had come and taken El in. Though he could tell them nothing, not knowing how to speak, the villagers helped and healed him. They gave him food and shelter. They taught him their ways, even how to communicate with words.

"El learned and grew and stayed amongst them, never forgetting their kindness. He learned what it was to hope, to love, to dream. He learned about how humans lived and died. And so, slowly, he came to understand them.

"Finally, he called to the Mother as he'd not been able to the first day. Proud, she took him from the village and made him once more what he'd been. But El didn't forget what he'd learned, or who had taught him; so he returned once more to the village, showed himself as the powerful god he was and then told them they were his Chosen.

"He lifted their land and other fertile places into the air

as a gift to them. He also gave them the ability to change the path the islands could take as well as a flying ship for each one. Before them, he proclaimed to the Mother and all the other gods these people were his and His alone."

"It must be wonderful!" Larana's awe pained him. "They get to travel and see the world without ever having to leave home." She looked at him, an expectant smile on her face. "Do you think we might meet some of them?"

"I doubt it." He tore his gaze away from the island. "The Chosen don't associate much with grubs." He suddenly wished there was no need for him to go to Caeldanage.

"Grubs?" Larana stared at him, her brow furrowed, almost tripping as she paid little attention to where she was going.

"Yes, grubs—or when they're feeling magnanimous, Landers. It's what they call those who live on the ground."

"And because of their homes is why they're called Flyers?"

Torren lied. "Yes." He then increased his pace to forestall any more questions.

By the next day, the buildings on top of the island and the tall spires beside them were visible. They drew the eyes as the island grew bigger by the hour.

As Torren and Larana crested a sizable hill, they discovered Caeldanage spread out before them. The city was several times larger than the floating island, yet the island's shadow ominously occluded a significant section of it from the sun. He frowned at the sight.

"It's huge!" Larana stared from the city to the island to Torren and back again. She pointed at the city. "Is it really full of people?"

He nodded, watching as she eagerly studied the city's high wall and what she could see of the crowded buildings behind it. The governor's fort rose as a domineering form from the city's center; yet it, too, looked less than what it ought, shrouded as it was by the island above.

"You'll need to be very careful here. You're going to see many things and people you've not seen before. They're not all good."

"All right."

He looked away as she gazed at him, her eyes full of trust and innocence. He knew it would all change soon enough.

By late afternoon, they reached the city gates. Larana's face was pale as she stared up at the towering three-story battlements. Guards stood at small booths just beyond the huge gates as well as on the ramparts. Their solemn, unwelcoming faces and the heavy presence of the island above pressed down on them with living force.

"Come on, let's find a place for the night." He pulled on her sleeve and entered the city. They hadn't gone far before she all but pressed herself against him. They'd come to a place vastly different from anything she'd ever known.

Strong smells assaulted their nostrils—old sweat, perfumes, animal droppings, food—all mixed together in a cacophony of scents that left one breathless. Stone and wooden buildings, some several stories high, rose on either side of them along narrow streets, closing them in. People moved back and forth, reminiscent of changing river currents. Occasionally, here and there could be seen a little green growing from a pot or planter; otherwise, everything else was wood, plaster, stone or flesh.

Larana tried to stare everywhere at once as he led them deeper into the city. Torren, however, had been here a number of times and had a particular destination in mind.

It'd been a couple of years since his last time through, but the city didn't look to have changed much—on the surface, anyway. The residents seemed more in a hurry than he remembered, the expressions on their faces less carefree. But then, with the weight of the island hanging over them, he couldn't blame them.

As they skirted a large bazaar filled with permanent stalls and colorful tents, Larana jumped up and down at his side, trying to get a good look at what there was to see. The scent of clustered hot bodies was strong, but so were those of frying oils, sweets and bread.

"Torren!"

He stopped as she pulled on his pack.

"Look! Are those...Are those...?" Her voice trembled in

her excitement. He turned to see what she was pointing at. "Are they Flyers?"

He froze at the question, but it was already too late. There they were, carefully walking amidst the stalls—the Chosen. People were giving them room, parting before them even as they ogled them. Their fair hair and skin, hairless faces, and their quiet beauty would have made it hard to mistake them for anyone else. Their short, layered robes and exposed legs also called out their foreignness. And if anyone still doubted who they were, the wings protruding grandly from their backs would have left no doubt whatsoever.

Torren felt his throat go dry. He hadn't seen any Chosen since...

"Let's go." Grabbing Larana's arm, he turned abruptly and merged into the flowing crowd. She didn't resist, though her surprise was clear.

"They're beautiful!" She struggled to keep up with him once he let go of her arm.

"I know what they are." He strove harder to get away from the area.

By the time he got them to their destination, his temples were throbbing, his body taut. Why here? Why now? After all this time.

Larana stopped beside him, panting, a questioning look on her face. "Torren?"

Without saying anything, he opened the well-worn door to the Wide Brim Inn.

The common room was large, big wooden tables and benches spread at odd angles throughout. Wide windows allowed in some of the outside light, showing sturdy rafters and a scarred but clean stone floor. A fireplace took up most of the left wall, its mantel holding bronze castings of the symbols of the nine gods. A broad flight of stairs took up the rest of that side.

Opposite the door stood an extensive bar, a wide selection of bottles set in niches behind it. Double swinging doors led to other, unseen rooms in the back.

A bell sounded as they opened the door, and they'd barely gotten inside when a portly man in a newly stained apron emerged from the back. The proprietor wiped his hands on it as he approached, a well-schooled welcoming smile on his face. A large, jagged scar followed his jawline, cutting a clear path through his full, peppered beard.

As soon as he saw them, his smile faltered, but his deep, penetrating brown eyes brightened.

"Torren?" The smile kicked back in with almost blinding force. "By the First Mother, it *is* you!"

He crossed the common room in big, bounding steps and, before Torren could stop him, clasped him in a back-breaking hug.

"Sal, please!" He suffered through the display as best he could, especially since the man pointedly ignored his plea and he was helpless to do anything about it.

After a moment, Sal pulled away and held him out by the shoulders so he could take a good look at him. He scanned him quickly up and down.

"It's been a while, hasn't it? I was starting to think you were avoiding me. But it's still good to see you."

Torren looked away, not having made up his mind before whether or not he'd have the time to call on his friend. "You, too."

"Still as grim as ever, though, I see." Sal gave him a raised brow.

He shrugged, thinking his friend a bit too merry. A stifled giggle sounded behind him.

"What's this?" Sal asked, taking a look past Torren's shoulder at his shy companion.

"That's Larana."

Sal's gaze shifted sideways to throw an inquisitive look in his direction as he added nothing else. Torren elected to ignore the hint.

"Still as tight-lipped as ever, too," Sal half-whispered out of the side of his mouth as he stepped around him to greet the girl. "Welcome, miss, to the Wide Brim Inn. Come, sit down, both of you."

He drew them both deeper into the room, steering them toward a table. "Are either of you thirsty, hungry?" He was headed back the way he'd come before receiving an answer. "I'll have something out for you in just a minute. Sit, make yourselves at home."

Torren wasn't sure if he was ready for Sal's generous

hospitality, but it didn't appear as if he'd get much of a choice. Giving in to the inevitable, he sat down as Larana surveyed the room, and rubbed at his temples. He wanted nothing more than to just get this business over with. The sooner he got out of Caeldanage the better.

Sal returned carrying a laden tray. Several faces peered through the kitchen doors. Torren straightened up, trying to ignore the discomfort lingering in his shoulders.

"Miss, the food is here." Sal beckoned Larana to join them as she gave a small wave at two women still peeking through the doors. Sal glanced at them, and they instantly disappeared. "Sorry about that," he said sheepishly. "But you know how the girls just love to look at you."

Torren gave him a sour look.

"Hey, if there was a way, I'd trade my ugly mug for yours in an instant. You've always been a very handsome fellow, aside from your nasty habit of grimacing and frowning all the time. I bet if you tried to smile more, you'd even give a Flyer a hard time."

His frown deepened. Sal laughed at his glowering expression and clapped him hard on the back. "Don't believe me, do you?" His eyes grew mischievous. "Miss, don't you think this man here strikes a handsome figure?"

Larana stopped in mid-chew on a seasoned bit of beef and seriously studied Torren. After a moment, she nodded slowly, her cheeks filling with color. Torren turned his scowl on her, and she quickly looked away.

"See? And just think, if you'd actually take my advice and lighten up your disposition, how would any woman be able to resist you?" Sal grinned as he spoke, smacking him on the back again.

He put up with the pounding, trying his best not to give Sal the deeper frown he'd be expecting. It was hard. "Don't you have something else you should be doing?"

Sal showed them his brilliant smile and sat down. "Nope. Your timing, as usual, my friend, is perfect. Most of the meal preparations are well on the way and being supervised by the cook. The crowds aren't due for a little while yet." He stretched, his eyes dancing, and swiped a bit of meat for himself. "I'm all yours."

"Great." The frown he had been trying so hard to hold back got the better of him.

Sal laughed then mercifully removed his attention from Torren to Larana. "Well, young miss, how are you liking the city so far?"

Torren grabbed a mug of ale off the tray. She hesitated before answering.

"It's so big."

Sal grinned. "First time here, I take it?"

She nodded. "Are they all this large?" she asked timidly.

"Some bigger, some smaller. We're quite proud of ours, though—it's one of the bigger ones on this side of the empire. If you're staying here long enough, perhaps I can arrange for a tour." He gave her a wink.

"That would be very nice."

"All right, then." Sal's attention returned to his friend. "So, Torren, what brings you up this way? Could it be you've finally decided to run this place with me, or might it be those rumors of work coming from up at the border?" He sent a quick glance in Larana's direction. "Or perhaps something else altogether?"

Torren took a deep drink of the dark ale before responding.

"Just the usual." He didn't elaborate. His friend would just have to wait for explanations at a more convenient time.

"I see." Sal sat back, taking the second mug of ale, leaving a smaller cup of watered wine for Larana. He took a deep draught then gave a long, appreciative belch.

"Have there been many rumors?" Torren asked as casually as he could.

Sal studied him from beneath bushy eyebrows and rubbed for a moment at his jagged scar. "Too many to figure out what's true or not, and definitely nothing to substantiate any of them."

Torren nodded and looked away, finally grabbing a piece of meat to chew on.

"The only thing worthy of note, as if you could have possibly missed it, was the arrival of the Flyers." The last was said in hushed tones.

Torren half-worried Larana would speak up and say or ask something about them again. Instead, she sat quietly sipping her wine, as if trying hard to stay out of the way.

"How long have they been here?"

"Oh, about four days or so." Sal shrugged. "There's almost as many rumors as to why they might be here as there are about what's going on up north. All I know for sure is they're not here for anything but business—and not trading business, at that, or else I'm sure they would have used one of their flying ships. Still, other than a select few coming down to the embassy or going over to see the governor, none have come down, which makes little sense.

"For some years now, there's been tension between us, which is easy, since no one is all that wise about the other, but there's more to it now. And no one's really talking about what's going on."

A feeling of foreboding swept through Torren. It was foolish, though, and he knew it. The two events couldn't possibly be connected.

"So, young miss, has this rascal told you yet how we met?"

Torren caught Larana throwing him a guarded look, and he groaned inwardly.

"No."

"Sal..." He threw a tone of warning on the name.

His friend ignored him. "Well, now, I guess it's going on about nine years now, wouldn't you say, Torren?" Sal smiled as his brow furrowed. "Still a pup when he joined the ranks of the mercenary force at Zellos." He leaned toward her. "It may be hard to believe, but the scowl you see right now was even more ingrained back then."

Larana's light eyes darted in Torren's direction and even more quickly away.

"Sal."

"Anyway, he was totally green but demanding to be allowed to fight. Didn't even have a concept of the rudimentaries. Almost as if he had a death wish."

Again a sharp glance. Torren felt like groaning out loud.

"So, though he wasn't too thrilled about it, I took him

under my wing and tried to teach him proper. Not that he was amenable to the idea, you understand. He's stubborn, this one."

Sighing, Torren stared at the floor, knowing there was no way Sal would stop until he'd had his say. Stubborn he had been, but in Sal he'd found his match.

"Good thing I taught him, too," Sal continued. "If I hadn't, he wouldn't have been able to help me keep this from going across my throat instead of my jaw." He ran his finger down the scar and grinned.

"It was then I decided I'd pushed my luck far enough and took on a slightly safer line of work. Tried talking him into joining me—he owns half the place as it is—but the wanderlust is too strong in him. Unfortunately, it also means I don't see him as often as I'd like." He smacked Torren hard on the back, almost knocking him off the bench.

Larana giggled as Torren threw his friend an evil look.

"Yes, it's always such a pleasure to see you, too."

Sal's deep laughter filled the room.

"Ah, my friend, how I've missed you!" He wiped at his eyes, still chuckling.

Larana worked up to a question. "What is it like to be a mercenary? Do they allow women to be them, too? Do you get to travel a lot? Is it hard?"

Sal's brow went up at the bombardment. "Whoa, whoa, one at a time, lass."

"Sorry."

"I take it you're considering some kind of career change?"

Her gaze flitted from Torren to her hands on her lap. "I don't know."

Sal threw him a grin. "Well, mercenary work is not for everyone. It's a hard, thankless life, actually. I doubt it would appeal much to you."

Larana's brows drew together. "Then why did you do it?"

"Ah, a fair question." Sal nodded slowly. "To be honest, when I was young, I wanted adventure, to see the world, and mercenary work was the fastest way to get it. No sitting around the blacksmith's shop for me, no, ma'am. I've gotten an eyeful, too, and seen and met a lot of strange folk.

"But there's an ugly side to it, too, and not everyone is cut out for that. Killing's never easy. This, here, is more of what life's about." Sal gestured at the room around them. "Instead of going to other places and people, I have them come to me.

"Now, if I could only get some other people to see that as well." He sent Torren a pointed look, which he studiously ignored.

The door opened, setting off the large bell attached to the door frame. Sal stood up.

"Welcome! Take a seat, friend, and I'll be right with you." He turned back. "Ah, looks as if work is here. Tell you what, when you're ready, go on up the stairs and take the last room at the end of the hall on the right. Make yourselves comfortable. We can talk again later."

He winked at Larana, as if telling her there'd be more fun to come then moved off to deal with the waiting customer.

Torren watched him go then stood up, grabbing his pack. Larana quickly popped a last piece of meat in her mouth before following suit. They took the stairs to the second-story landing and followed the wide corridor to the last room on the right. He opened the door then stepped aside so Larana could go in.

The room was neat and clean, with two large beds with straw-filled mattresses and a sturdy dresser with a pitcher and basin. A large window looked out onto the slowly darkening street.

He watched Larana from the doorway as she inspected the room. They were here, she was safe, it was time for him to do what he'd come here for in the first place.

"I've got to go arrange a couple of things before nightfall. Wait here for me."

He tensed as a look of panic crossed her face, though she quickly brought it under control. Did she somehow suspect what he was up to?

"Will you be long?" she asked.

Torren couldn't meet her trusting gaze. "Not long."

"All right."

He turned away, closing the door, knowing this was the last time they'd see each other. He was surprised when he found he was feeling regret at the thought. Shaking his head at his foolishness, he made his way back downstairs.

Four more people had entered the inn looking for drinks and dinner. The two women who had been in the kitchen were now waiting tables. Sal had taken his place behind the bar, a captain at his ship's wheel. Torren headed straight for him.

"Is the room all right?" Sal asked brightly when he saw him.

"I need to talk to you," he said. "And I don't have a lot of time."

Sal stared at him, taken aback, but nodded. He set down the mug he'd been filling and signaled to one of the girls. "I'll be back in a minute."

He led Torren through the double doors into the kitchen. They passed a large wood-burning stove, where a man and an older woman tended the food, and went into a storage room. Sal slid the door mostly closed, trapping them in shadow except for a small slit of light. He turned to Torren, his face full of apprehension.

"What is it? Trouble?"

Now that the moment was here, Torren found he was at a loss how to begin. "I—I have to go. But I can't take Larana with me." He reached into one of the side pockets of his pack and brought out a small bag holding most of the money he had on him. "Her family is dead, killed by bandits." He handed the bag over. "There should be enough here to pay for your trouble as well as for an intermediary to find her a husband or a home. If anything's left, you can give it to her or keep it, whatever you see fit."

He turned to leave.

"Torren, why?" Sal reached out, his arm blocking his way. "Taking care of the girl is generous, and I can understand it, but why are you leaving this way? This isn't like you."

He only shook his head. "It can't be helped, and I can't explain it now. What time do they close the gates?"

Sal frowned, examining his face intently. "I doubt you'll reach them in time. Since the Flyers came, they've been more prompt than usual and have been closing the doors as soon as it grows dark. What are you running away from? Does she even know you're going?"

He heard a trace of anger building in Sal's voice. It was prompted by his friend's strong sense of protectiveness. The same one that had made him help Torren all those years ago, despite the fact his help hadn't been wanted. It was why he was counting on him to help Larana now.

"I can't explain. Just do this for me. Please." Without waiting for a reply, he slid the storage room door open and ducked beneath Sal's arm to get out. He turned in the opposite direction to the common room and left the inn by the rear exit.

Torren was short of breath as he stepped out into the cooler air of the alley. He looked up, but stopped himself before he could look for the floating island. He'd done it; he was rid of her. Now all he needed to do was get out of town before the gates closed.

# CHAPTER 7

THE STREETS HAD THINNED OF TRAFFIC WITH THE COM-
ing of darkness, but the press of bodies around the bazaar
was heavier than before. His mood growing dark, Torren
pushed and shoved, trying to get through as fast as possible. He rushed along the streets once the way was clear
again and headed for the southern gate of the city, as it
was the closest. He glanced over his shoulder at the island
pressing down on him from above. If only *they* hadn't been
here. He put on a burst of speed.

He came around the last corner before the gate, and the
hope that he'd make it was snuffed out like a candle—five
guards were dropping the heavy bar into place. No one
would be coming in or out of the city tonight, and the city
guard would as soon gut you as listen to why you might
need to do either.

Disgusted, and with a growing feeling of unease, he
turned away before the soldiers noticed his presence and
demanded to know what he was doing there.

To return to the Wide Brim Inn was out of the question.
By now, Larana would have asked after him, and Sal
would have been forced to tell her he was gone. Instead,
he hunted down one of the cheaper, seedier taverns on
this side of the city. He still wasn't sure what had possessed him to give Sal most of his money to care for the
girl, but what was done was done. He'd owed her nothing,
had even gone out of his way to help her when it might
have been more prudent to do otherwise, yet he'd not been
able to help himself in the end. It was good he'd gotten rid
of her before the lines became any more blurred.

The Stag's Horn was a world apart from the Wide Brim Inn. Smoky and dingy, the interior looked as if it had been years since it'd seen better days. The stink of sour ale mixed with vomit lingered beneath the smells of sweat and cooking meat fat. Despite all this, the place was already half-full.

Torren grabbed a small table near the bar, studying the place's clientele from the corner of his eye. Caravan guards, a few low-end merchants and laborers comprised most of the lot. Here and there even seedier characters sat, either avidly avoiding eye contact with anyone or studying the crowd for future business.

The woman who eventually brought him some ale and a bowl of fatty stew looked as if for her, too, it'd been years since she'd seen better days. He paid what he owed and also got a space upstairs while he held her fleeting attention. He wasn't pleased at his current lack of funds, but he'd been worse off before.

He did have more, not being one to squander his wages as so many others in his business did on women, drinking and gambling; but it was stashed in different cities and investments, all too far away to do him any good at the moment. Perhaps he could speak to some of the men here and hook himself up to a caravan in the morning, earn some pay as he continued north.

Most of the conversations he could overhear were the usual grumbling about wages, unfriendly tavern wenches, bets lost in local games. Yet another topic seemed to weigh heavily on a lot of their minds, the resentment and anger in the words more pronounced than usual.

"I'm telling you, someone's got to do something about them. Just because they have wings and those stupid islands of theirs doesn't mean they have a right to put the things wherever they want." A sour-looking blacksmith, still wearing his apron, chugged back a swig of ale. "The sunshine disappears about midway through the day. It's depressing! The wife is getting moodier than ever. She says they're keeping her bread from rising."

"And just what do you expect anyone to do, Lucius? How are you expecting anyone to get up there?" His companion laughed, yellowed teeth gleaming in the bad light.

Lucius's face turned red. "That's the emperor's problem, not mine." He chugged another swallow.

"But how is he supposed to handle their powers? I know they look normal enough, except for the wings and all, but I heard they can put a spell on you if you meet their eyes..." The new speaker jabbed out two of his fingers like arrows from a bow. "...and make you do their bidding. It's why they don't send more goods by caravan—the Flyers bewitched the politicos into choosing them instead of proper men."

Torren would have laughed at the last, the concept ludicrous; but unfortunately, a lot worse things were believed of the Chosen—mind control, causing sickness, immortality. A grisly tale heard from some mercenaries during one of his first campaigns told of how Flyers would search for those of great skill and power then claim them in the night to eat them, sucking up those very traits. It was amazing how these tales clashed with others he'd heard proclaiming the Chosen as servants of the gods, their floating cities a kind soul's ultimate resting place. It was almost eerie how they could be loved, envied and hated, all at the same time.

He was forcing some of the overcooked fare down his throat with the help of the bitter ale when the door to the tavern opened. Now, these fellows looked out-of-place—a powerful man in black leather armor with two others similarly dressed trailing behind him. The leader's manner was aloof, as if nothing here could touch him—not the stench, the cheap wine, the worthless customers. Their dark armor looked well-oiled and cared-for. Their scabbards were of obvious quality as well. People like them would never be caught in a place like this, not unless they wanted something.

With his two companions in tow, the stranger headed straight for the bar. The whole room grew quiet at their entrance. Some must have thought it didn't bode well, for a couple of patrons were already trying to make their way unobtrusively toward the door.

"Good people," the stranger said, turning toward the crowd while flashing them a bright smile, "I'm looking to hire a few good men for a small job this evening." His cold eyes swept the tavern. "I only need five or six individuals who know how to be discreet."

As he spoke, one of his companions took out a pouch and let it land with a heavy jingling of coins on the counter.

"Might there be any of you here fitting this description?"

The question couldn't have been put more sweetly. A number of the men in the room stared lustily at the pouch.

"What might this here be about?" asked a voice from the back.

Torren listened intently as he pretended a lack of interest. Unless his business wasn't slightly on the shady side, this wouldn't be the kind of place one of that type would be coming to enlist help.

At the question, the stranger lost his smile and looked saddened. "I can't go into a great deal of detail now, but let's just say I'm looking for some courageous men to help my niece disassociate herself from some, shall we say, disreputable people."

Torren's eyes narrowed. He wanted help to rescue his niece? Help from people like these? That was more the province of the city guard or even well-reputed mercenaries, not riffraff.

The people in the room exchanged looks, many of their thoughts obviously going down the same road as his. No one made a move toward the bar.

"Come on. Surely, a few of you would be willing to help her. I fear for her life, so speed and stealth are of the essence. And everyone knows the guard is not very good at either."

This elicited a few laughs from the crowd. Whispers ran through the room, but still no takers stepped forward.

"No one?"

The man behind the speaker took out another pouch and dropped it beside the first. This one slipped open, spilling out several gold royals and a platinum. In-drawn breaths whispered across the room.

"For a piece of that, I'm your man!" What Torren guessed to be a caravan guard stepped forward. His eyes shone with barely restrained greed.

"That's the spirit! So, who else is brave enough to join my cause?" Three more men stood and shuffled forward.

That this group was up to no good was now obvious. Torren had no intention of getting involved. Let this be someone else's problem, whatever it was.

Still, as the volunteers gathered about the suspicious men at the end of the bar, he realized his ale cup was empty. Not looking at any of them and putting a slightly drunken swagger to his step, he made his way to the bar as well. Leaning up against it, not looking at the group nearby, he kept his ears open.

"...she's young, just a lass. All you have to do is look for her in the guest rooms..."

"Hey, you!"

Torren jerked around and pretended to lose his balance as someone tapped him hard on the shoulder.

"Whoops!" He giggled as if he were having a hard time staying on his feet. Inside, he felt terribly cold.

The leather-armored man stared at him in disgust. "Go on back to your table. You're drunk, and we have no need of those who can't hold their liquor."

He gave Torren a push. Hopping on one foot to stay upright, Torren giggled again before making his way back to his table. As soon as the man saw him go, he turned away and went back to the ongoing discussions.

Torren sat down and giggled once more for effect then went still. He stared at the scarred tabletop, the stew he'd eaten turning to stones in his stomach. It all had to be a coincidence—it had to be! There were hundreds, if not thousands, of young women in this city; and any number of them might be lodged at an inn. Yet something inside him insisted there was no mistake.

These men were the same ones who'd killed Larana's aunt and uncle, torched their home and chased her through the forest. And mere bandits they were not. What could they possibly want with a farm girl? How had they even found her again?

He shook his head. What did it matter? He owed her nothing, nothing at all. But what about Sal? These men were going to raid his inn. Whether he liked it or not, he owed the man his life and more. He'd have to at least warn him they were coming. He could do that without seeing Larana.

Torren glanced up at the men still piled at the bar. None of them was paying any attention to him, too engrossed in their current conversation. So, he slumped over, picked up his pack and pretended to drag himself upstairs. As soon as he was out of sight, he straightened and put his pack back on.

At both ends of the narrow hallway were shuttered windows. Seeing no one else about, he headed to the one in the far back. He almost smiled as he discovered the glass had been broken previously and no one had bothered to replace it.

Swinging open the shutters, he spotted an awning not far below. He suspected it had been used before for the same purpose he was about to put it to now. He slipped his pack out the window then followed it.

Clinging to the awning's edge, he dangled for a moment then let go, to be swallowed by the darkness of the alley. His senses primed, he sneaked down to where it opened out into the street, the smell wrapping around him even worse than what he'd found indoors. He looked back toward the tavern and spotted two men on horses; they wore the same well-cared-for black armor as the others.

Skirting from shadow to shadow, Torren hid from them as he hurried down the street. As soon as he was out of sight, he took off at a full run back toward the Wide Brim Inn and the girl he'd thought he'd not have cause to see again.

By this time of night, most of the streets were deserted. He stuck to the main roads, not wanting to pick up other kinds of trouble despite the risk he might instead encounter some of the Guard. The street predators would be out in force by now, looking for drunken pickings; and he had no desire or time to deal with them.

When he reached the back of the inn, he leaned forward with hands on thighs for a moment, trying to catch his breath, his side aching and his pack feeling akin to a mountain on his back. He knew he had only bought a few minutes. The men in armor were probably already on their way. Straightening up again, he took out the knife from his boot and used the pommel to strike the door.

The old woman he had seen earlier eventually opened it a crack and peered out. He instantly pushed on it, driving her back with a gasp.

"Get Sal, get him now!"

The woman gasped again, seeing the steel of his knife glinting in the light, and scurried through the kitchen to the common room. Torren put the knife away and dropped his pack on the floor before leaning against the wall. He closed his eyes for a minute. His heart hammered in his chest, but he knew it was from more than exertion.

Sal rushed into the kitchen, his expression angry until he got a look at who'd frightened his cook. "Torren!"

He opened his eyes and pushed away from the wall. "Trouble's coming."

"What?" Sal stared at him, perplexed.

"At the Stag's Horn, some military types were using gold to hire anyone willing. They're coming here."

Sal cursed under his breath. "What do they want?"

He shook his head. He didn't enjoy keeping things from Sal, but it would be better this way. "All I know is they're coming here."

Sal nodded, never showing any doubt of his word.

"Kyran," he shouted over his shoulder.

The man Torren had seen earlier in the kitchen appeared, an antagonistic look on his face whenever he glanced in his direction.

"I need for you to go find the Guard. Tell them we have a riot, anything you have to, but make them come back with you. Do it now."

Kyran nodded, surprised, and slipped past Torren, after one last glare, into the night. Sal watched him then turned to go back into the common room. Picking up his pack, Torren followed. Sal went behind the bar and, from a shelf near the bottom, removed a long object wrapped in cloth.

"Didn't think I'd ever have need of this again, but thought I'd keep it handy just in case. There's nothing like a little excitement to spice up life." He unwrapped the bundle and revealed a sheathed broadsword. He pulled the blade and laid it down on top of the bar. He stared out at the ten or so customers still there.

"Sorry, folks. Looks as if a bit of trouble is coming our way, so I'm going to have to close early." A grin spread across his face. "You can come back and see what's left tomorrow."

The two barmaids traded startled glances and so did the customers.

Sal held up his sword when no one made a move to go. "Are you all deaf? Go on! Go! And if you run into any of the Guard on the way, send them here quick."

Two of the customers headed for the stairs; the rest went quickly out the front door.

"Mila, Sheree, go on home. You can clean up tomorrow."

The two girls promptly fled into the kitchen.

Torren watched all this with detachment, his mind occupied with other things. All at once he looked up at Sal and asked, "Where's Larana?"

The half-grin died on Sal's face. His friend stared at him with something close to reproach. "She's still where you left her."

Not taking the time to think about exactly what Sal meant, he headed for the stairs. He'd given warning, and the Guard would soon be here; but it was still no guarantee those men wouldn't get what they were looking for.

And even if they didn't, there'd be questions afterwards. If any of the hired men were caught, they'd talk about their leader's alleged "niece." With all the money changing hands, the Guard wouldn't believe Larana when she told them she knew nothing about what they'd been told.

Torren took the steps two at a time.

When he reached the end of the hall, he opened the door without knocking. A candle burned low on the dresser, throwing dancing shadows on the walls. Larana lay on one of the two cots, her back to him. His brows drew together for a moment as he spotted the money pouch he'd given Sal sitting in the middle of the floor, its contents partially scattered about. It reminded him of the similar scene at another tavern not an hour before.

He set his pack down and knelt to gather the coins. The

two situations were in no way similar, but he still couldn't shake the feeling they'd been very much the same kind of payment. Putting the pouch away, he glanced once more at Larana's still form.

She was stretched out on the cot diagonally, her shoes still on her feet. It was almost as if sleep had caught her unawares. He knew from seeing the money Sal had spoken to her of his leaving. What would she make of his sudden return? Hesitantly, he reached out to shake her arm.

"Larana."

He took a step back, startled, as his touch brought her instantly awake. She whirled upright, red, swollen eyes wide and staring in a shocked face.

They were immobile for what felt like an eternity, until her eyes flashed with sudden temper.

"You *left* me! You didn't even say goodbye. Why? I thought...I thought we..." Her anger died as he continued looking at her in impassive silence, replaced by misery and self-doubt. Tears welled as she abruptly rushed off the bed and threw herself at him. "Torren!"

He caught her easily, startling himself by doing so. She sobbed against his chest, pounding him with her small fists.

"You shouldn't have done it. I don't want your money! You're the only one who understands, don't you know that? You could have told me...You could have told me..." Her words turned incomprehensible as her warm tears soaked through to his skin.

Hesitantly, calling himself a fool, he placed his arms comfortingly around her. Guilt infused him at what he'd done, but what choice had he had? Now things were about to get even more complicated.

"Larana," he said gently. He pried her carefully off him and stared down into her wet face. "We have to go."

She stared at him in confusion, alternately trying to wipe her face with her sleeve and control a sudden bout of hiccups.

"I don't understand."

"Those men are back. We have to go."

Though he could tell his words only made her more confused, she nodded.

"I'm ready."

Taking her at her word, he picked up his pack and headed for the door. She was right behind him. They met Sal in the hallway as he pounded on his guests' doors to rouse them.

"Ah, you've got her. Good."

Startled customers peered out into the hallway.

"Good people," Sal advised, "trouble is coming our way. I've sent for the Guard, but they may or may not arrive in time to avert it. I suggest you get dressed and leave, or stay, if you desire, and prepare yourselves for a possible fight."

He turned away and followed Torren and Larana down the stairs.

"Are you staying?" he asked, his grin back on his face as the sound of rushing footsteps sounded above.

Torren glanced at Larana and answered, "We can't."

Sal studied him intently. "Why do I get the feeling there's more going on here than you're telling me? It's not like you to run from a fight."

He looked away. "There's a lot I don't know. I'll have to try to explain it to you later."

It was the best he could do. He hoped Sal would accept it as enough for now.

"I'll hold you to it, and I don't want to have to wait years, either." He slapped him lightly on the back. "If you think you have time, take some food with you." His eyes softened as they turned toward Larana. "I expect you to take care of this stubborn man, miss, at least until I see the two of you again."

Torren blinked in amazement, though Larana nodded slowly, looking happier than moments before.

Sal glanced over his shoulder at the stairs for a moment before reaching for a couple of cloaks hanging on pegs by the stairs.

"Here, you'd better take these. They should help keep you from being noticed." He handed the garments over, grinning at Torren's expression. He steered them toward the kitchen. "Take care."

70

Once through the doors, Torren regained his mental balance and hurried on. He grabbed a couple of loaves of bread, a chunk of cheese and some smoked fish hanging from the ceiling. Larana grabbed a couple of onions and peppers as well as some fruit from a bowl by the door. He made no comment as they dumped their assorted booty into a sack he fetched from the storage room.

Making sure she was wearing the overly large cloak Sal had given her and that she stood behind him, he opened the door to the alley and slipped outside. Handing Larana the sack of food, he unsheathed his shortsword and led the way down the dark street. They traveled from shadow to shadow. All three moons shone above them, a strange counterpoint to the silhouette of the island looming over them. The governor's fortress rose imperiously between the two.

Once he made it close to the northern gates, Torren stopped and peeked around the corner at the gate's guard station. Light streamed from the open door into the street. Larana lightly bumped into him and mumbled an apology before leaning against the wall. He glanced back at her but could see little beneath the cloak's deep cowl. From her slumped posture, though, it was easy to tell the girl was dead on her feet. He didn't feel much better. It had been a long day for both of them.

"We're only going a little farther." He took her unresisting arm and led her back the way they'd come for a block, out of sight of the guard station, before crossing over to the other side of the main thoroughfare. Against the right side of the city wall, large, narrow barns with gated stalls stood next to several closed storage areas. Everything was quiet.

Still leading Larana by the arm, Torren sneaked into the closest barn through a partially open door. The scent of hay, animals and manure swept over them as they entered. A soft whinny disrupted the quiet but that was all.

Torren groped about in the dark until he found a ladder leading up into the hayloft. He prodded Larana to go on up. The girl obeyed, almost slipping twice on the rungs. As soon as he'd gotten her and himself to the top, he led her to the far corner, where she collapsed into a heap and didn't get up again.

Torren knelt beside her, moved back her cowl to check on her and found her fast asleep. He studied what little he could see of her long face for a moment then pulled some loose hay down to cover them both. He fell asleep almost as fast as she had.

# CHAPTER 8

ACTIVITY BELOW SLOWLY BROUGHT TORREN AWAKE. HE didn't open his eyes immediately, his body warm and strangely comfortable. He was tempted to allow himself to fall back asleep, until the events of the night before trickled back into his consciousness. Now totally alert, he realized Larana was snuggled up against him.

Apprehension mixed with reluctance swept through him as he gently pulled away. Strangely, he found himself wondering what Sal would have made of the scene. He hoped his friend had fared well in the night.

"Larana," he whispered close to her ear. She mumbled a few unintelligible words and curled up into a ball. "It's time to get up."

He shook her lightly.

Larana grudgingly opened her eyes then suddenly sat up in alarm.

"It's all right. We're safe."

She looked about their current refuge, her expression showing she possessed no recollection of having come there. Sounds drifted up from below.

"I have to leave for a little while," he told her.

Her face instantly filled with panic. "No!"

Torren felt his own harden though a pang of guilt rang inside him—after the last time she'd have no reason to believe he'd be back.

"I have to arrange for us to get out of the city. After not finding you last night, it's very likely those men will have lookouts posted at the gates."

"Why are they looking for me?"

73

He shook his head. "I don't know. You may have seen something you shouldn't have." He tried hard to smile. "Maybe it has something to do with your strange gift."

Larana's brow furrowed as she stared at her lap, absorbing all he'd said and mulling it over. "But no one knows about my gift. It's a secret. And I don't remember seeing anything."

Her sky-blue eyes met his.

"It could be you saw something you don't remember seeing, or it's something they only think you saw. As for the other, there's a chance your aunt or uncle might have told someone. Your aunt came and woke you, as if she were going to show you to them, before she asked you to run."

Larana shook her head. "No. They...They constantly asked me to hide it. They wouldn't have told anyone." She stared at him, confused.

"We can discuss it some more later," he said, pushing the subject aside. "Right now, our priority is getting ourselves out of here."

He moved to the edge of the loft and looked down to make sure no one was watching, He deftly descended the ladder and sneaked out. Finding a shaded corner, he took off his cloak, shook the hay out of it then brushed off his clothes and hair and headed off toward the storage area.

After a few inquiries, as the sun rose oozing amber on the horizon, he was able to find the factor handling a caravan presently being readied to leave town. Some quick negotiations later, he returned to the stables and climbed back into the loft.

Larana, in the back corner with an apple in her hand, stopped in mid-bite, looking like a frightened child. As soon as she realized who it was, however, her face blossomed in relief and with something akin to joy.

Torren frowned, one part of him bothered by her reaction and dependency while another knew only too well what it was like to have no one to rely on.

"You can come down now. Though, until we're out the gates with the caravan, it'd be best if you showed yourself outside as little as possible."

"You bought us passage?"

"No," he said, "I got us work. I'll be helping guard the caravan, and you'll be doing odd jobs as well as helping with the mules and horses and assisting the cook."

"Good."

He was taken aback a little by her unexpected reaction. He'd half-expected she'd balk at having to work for their passage. Now he hoped her handling of Gimmel's mule hadn't been a fluke.

"I gave them our real names to keep things simple, but I've told them you're my sister, and we're heading north to visit a sick relative."

"But—"

He cut her off. "It gives us a reason to be heading toward the border, and it will make it harder for us to be found by those men if they question the caravan later."

He watched her give this some thought. After a moment she nodded, her face clearing.

Following a hurried breakfast, they gathered their belongings and headed for the ladder. As Larana started down, Torren leaned forward and removed a piece of straw from her hair. She froze, staring at him, and then gave him a dazzling smile, her cheeks coloring ever so slightly.

"Thanks."

He only grunted a reply, not sure why he'd done it and also amazed how hard his heart was pounding because she'd smiled at him so brightly. He swore to himself that, as soon as they lost these men, he had to find someplace safe to leave her.

With her hood safely over her head and hiding her face, he took Larana to meet the caravan master. After that, Torren pitched in with the loading of goods while she followed one of the livestock handlers to help harness the last of the mule teams.

Soon, all the wagons stood hitched, the goods were secured and everyone was ready to leave. Torren made sure Larana and their possessions were stowed in the back of the cook wagon then took his place in the line of guards.

The group left the loading area and headed toward the open city gates. The guards at the gate station only gave them a cursory glance after checking the caravan master's papers. Torren spotted two men on the other side of the gates sporting dark armor, surveying with interest everyone who left the city. Who were these people?

The morning offered no answers but also passed without incident.

At the midday rest period, he spotted Larana, without her cloak, helping water the mules and horses; he knew there was no help for her not wearing it when she was working. The caravan got moving again within the hour; and though he'd been keeping careful watch, he had spotted no signs of pursuit.

The caravan was of mid-size, composed of fifteen wagons drawn by four mules each. Two of the wagons were for food and supplies, one was a cook wagon, another a sleeping wagon for the merchant and the master; the rest were filled to the brim with goods from the south to sell to the northern regions and perhaps even Galt. The few horses brought along were for the merchant, the master and the head of security. Each wagon had a driver, and each driver was either part of the security force or a noncombatant with multiple jobs on the trip. Another ten armed men, including Torren, walked on either side of the line, providing added security.

Though they were traveling on a Grand Highway and small garrisons would dot the way until they reached the pass, the large amount of security showed how much the rumors from the north were affecting those coming from the south. Even the wages they were paying for this trip were higher than normal. Not that he was complaining on that account.

The caravan came to a stop an hour or so before sundown, pulling off to the side of the road. Immediately, everyone pitched in to set up camp, the master walking up and down issuing orders. Torren helped unhitch the mules and tether them in a line. The tree line in this area had been pushed back some time ago, and dark rings around small pits in the ground showed the site had been used many times by other caravans.

Larana and several of the others took feed from one of the supply wagons for the mules and horses. Others scavenged wood from the other side of the road and piled it into the pits. As the sky darkened, the fires provided the group with some light.

Not long after, Larana and the cooks came by with heaping bowls of meat and vegetables as well as cups of watered wine. She ran up, almost tripping, and with an embarrassed smile gave Torren his. Watching her run back, half-amused despite himself, he picked a spot by one of the fires farthest from the road and sat down. She was able to join him not long after. She settled down with a satisfied smile and dug in.

He watched her out of the corner of his eye, enjoying the heat from the fire as it mixed with the light, cooling breeze of the deepening evening.

"Have they been keeping you busy?"

She glanced over at him and nodded. "The cook had us cut vegetables and peel potatoes. Normally, it's pretty easy work, but I've not done it in a moving wagon before." She showed him where she'd accidentally nicked one of her fingers. "Everyone is really nice, though, and some of them have traveled all over."

The fire shone brightly in her eyes.

"Just be careful what you tell them," he warned. "What they don't know they can't pass on to others."

Her expression sobered. "You...You think they'll still keep looking for me?" Her question was spoken in less than a whisper.

He answered her as frankly as he could. "They've gone through too much trouble so far for me to think they'll stop."

She looked away, concentrating on the fire.

"Ah, there you two are."

Torren turned toward the voice and found the master ambling toward them. He gave them both a half-grin.

"Been hearing good things about you two. Heard you've been doing your proper share. Also seems the young lady is quite good with the animals. Hope you keep up the good work."

"Thank you," Torren replied. Larana nodded her concurrence, a shy smile on her lips.

"Yes, thank you."

"Well, I'll leave you to turn in. We should be getting a

pretty early start in the morning. Goodnight." He nodded to them and headed off toward one of the other fires.

# CHAPTER 9

THE GROUP HAD NO PROBLEM THAT NIGHT OR THE NEXT day. On their third day out of Caeldanage, they reached a small garrison town much the same as the one where Torren and Larana had met Gimmel and his family. The caravan converged on the town's inn, filling it to more than capacity.

Almost as if following a ritual, the caravan master, the merchant and the security chief all congregated around the innkeeper and questioned him for information on what they might expect nearer the border. Torren stuck close, trying to overhear; and a lot of what he heard was much the same as the rumors already circulating in the south. Even this far north, no one appeared to have any real idea what was going on, aside from the confirmed strange movements of men and supplies.

But even those didn't make clear which side was responsible. The innkeeper did add that the men of the garrison were drilling more than usual; and they, like everyone else, were tenser than normal. None of this eased the nerves of the merchant at all.

"It just can't be all true, really," he said. "I've been trading with Galt for half my life, and while they're a strange bunch, they don't look for trouble more than anyone else. It can't be them that's causing all this turmoil."

He didn't look convinced of his own conclusions.

"Who knows what motivates nomads?" The innkeeper polished the table he was sharing with the three men. "They worship the same gods we do and then some, but not the way we do it. You can't trust people like that. Flyers are just as bad, if not worse, sticking to only one god as if only theirs was important."

The merchant suddenly looked troubled. "Don't at least two of their islands cross over the border this time of year?"

"Yes, I believe that's right," answered the innkeeper. "And there's one over in Caeldanage, I heard."

"No telling which way the Chosen would go if war did break out," the security chief mused. "They might be just as happy if we all killed each other and they got to take over everything left."

Torren stopped listening as the conversation devolved into a general gripe session, instead nursing a mug of ale at the corner table, pondering the lack of information and the growing paranoia.

"Well, sir, might you be wanting a refill?"

He glanced up, not having noticed the woman approaching. "Yes, thank you."

She looked to be only a few years older than he and, from the cut of her bodice, knew how to display her finest attributes to their fullest potential.

Rather than take his glass to fill it, the barmaid leaned toward him, exposing even more of herself to his gaze. During the whole operation, her hungry eyes never left his.

"You're different from the kind we normally see around here." She roamed her eyes over the rest of him with definite approval.

He had a hard time breathing for a moment, slightly overwhelmed by the view. He'd been with women, though rarely, and only when his urges couldn't be appeased any other way. The thought of an accidental child resulting from a mild dalliance bothered him more than he was usually willing to admit. And even if he were so inclined, this time he wasn't alone.

He'd have to put a stop to this before she got the wrong impression.

"Yes, well—" He didn't get any further, as suddenly Larana came at him from the side and wrapped her arms around his neck.

"Brother! I've been looking for you everywhere."

He threw her an askew glance, wondering what she was up to. It was almost as if he wasn't there, though, as she smiled at the barmaid.

"Hello."

"Uh, hello." The woman didn't seem quite sure what to make of this.

"Keri!" The inn's owner waved at her. "Our guests need refreshments over here."

"Uh, excuse me." The barmaid threw Larana one last puzzled look and went to the other table. The moment she turned away, Larana let go of him.

"What was that about?" He rubbed his neck where she'd held onto him just a little too tightly.

She turned to look at him, all innocence and smiles. "Nothing. She just didn't seem like someone we want to get to know."

With no more of an explanation, she skipped away to talk to the cook.

Shaking his head, Torren turned back to his drink. As he sat there, he gradually became uneasy. He studied the innkeeper, his helpers, anyone he didn't know with increasing suspicion. Too many unexplained things had happened of late, and so much of the information he possessed was vague. It felt as if he were wading through a quagmire, unknown pits waiting for him to fall into them.

The temptation to just up and leave and get lost in the forest nagged at him, but he couldn't make himself give in to it. He had no obligations to this girl. He'd already done more than his share. Yes, she was an innocent, but so much trouble clung to her. Right now, she had a job, a place she belonged. It would be so easy just to slip away.

"Torren?"

He started, surprised as he realized Larana was next to him again. Her large eyes searched his, worry tinting her features.

"Is everything okay? I didn't make you mad before, did I?" she asked.

He felt suddenly ashamed for his previous thoughts. "No, everything's fine. Why?"

She looked away, her cheeks coloring slightly. "You... You were frowning a lot harder than usual."

She shyly glanced at him, as if trying to gauge his reaction.

He forced himself to smile. "Great, now you're starting to sound like Sal."

"He really cares for you."

Torren stared, realizing she was serious. "I suppose so. We worked together for a number of years."

Sal was also one of the few people he'd ever allowed himself to get close to.

"Do you think he's all right?" Larana looked as if she were half-afraid to hear the answer.

"Sal's a survivor," he told her, "and he loves a good fight. I'm sure he's fine."

That evening, everyone in the caravan got the luxury of taking a warm bath. Though it should have relaxed him enough so he could fall easily to sleep, Torren lay awake.

Larana was sharing a room with the cook while he stayed in another with some of the men. His feelings of unease didn't leave him, but he couldn't pin them to any particular thing. He was finding it bothered him not to be where he could keep an eye on the girl. Anything could happen to her here, and he wouldn't know it. Only a freak streak of luck had enabled him to keep her out of the hands of those who'd hunted her over in Caeldanage.

When he finally did fall asleep, his dreams were filled with moving shadows and a sense of unseen dangers.

In the morning, after what little sleep he was actually able to manage, Torren felt even more tired than he would have if he hadn't slept at all. To his astonishment, when he made his way downstairs, he found the grinding in his stomach loosening as he caught sight of Larana bouncing around the common room.

"Good morning!" She gave him a bright smile.

"Good morning." He sat down at one of the tables and let her serve him breakfast. "Been up long?"

"Not long. I was helping Bess get things ready."

"Larana, let's go." The cook came out of the inn's kitchen, a number of empty bags thrown over her shoulder.

"Coming!"

He reached out for her, grabbing her arm. She looked back at him, shocked.

"Where are you going?" He forced his hand to relax when she flinched in pain.

"Cook wants me to help her pick up supplies from the general store. They were going to open up early for us. That's all right, isn't it?"

He felt a tingle of fear come from her where he touched her skin. He let go of her arm, feeling like an idiot for making her feel that way. "Yes, it's fine. Go ahead."

Larana stared at him a moment longer, unsure, and then rushed to catch up with the cook, almost tripping in her hurry.

He gulped down the rest of his meal, not tasting much of it. Calling himself an utter fool, he stood up to follow them. Nothing would happen to her here. By the gods, there was a garrison across the way! His steps didn't slow.

The sun was just cresting the horizon in a bright orange ball as he stepped out of the inn. Briskly, he turned left and headed off in the direction of the general store. When he got there, he found one of the caravan's mules tied to the post outside, a number of empty saddlebags draped over its back.

Torren's brows drew together as he spotted a horse tied to the other post. The well-oiled saddle and the high quality of the animal screamed it wasn't one of theirs. What was it doing here? Only the caravan had taken rooms at the inn last night.

He turned where he stood, for the first time taking a good look at his surroundings.

Back toward the inn, voices rose and fell as the caravan crew started the slow process of getting ready to move out. The garrison appeared quiet, though he could see several guards stationed on the short watchtower's roof. The small market area beside it was empty. The only thing out of the ordinary was the well-bred horse.

His hand itched, hinting he should have had the sense to retrieve his sword before coming out here. His knife was still in his boot and would do in a pinch, but he would have felt better with the familiar weight of his blade at his side.

Stepping under the store's awning, he opened the door. Larana's easy laughter brought him to a stop before he actually stepped inside.

"You're going to have to show me how you do that!" a robust voice said from the back. "I think you'll be coming with me when I do my shopping from now on."

A moment later, he spotted the cook and Larana as they came down the aisle toward the exit.

"Torren!" Larana stared at him in mild astonishment.

"Ah, just what we need, a sturdy man," the cook said, grinning. "You're just in time to help us load the mule."

Without waiting for a reply, she handed him the filled sacks she was carrying and returned inside for more.

"Torren, is something wrong?" Larana asked quietly.

He forced himself to stop trying to spot the horse's owner inside the store and turned to look at her. "Everything's fine. Just making sure it stays that way."

Her eyes seemed troubled, but she said nothing further. Instead, she helped him pack the supplies away on the mule.

On the way back to join the rest of the caravan, Torren glanced back once but still saw no sign of the horse's owner. Once the caravan was ready and had set off, however, he noticed the horse was no longer anywhere in sight.

The day was pleasant, non-threatening banks of clouds cutting back the heat of the day and a cool breeze whispering down the road. He noticed none of it. His gaze darted from tree to tree, looking for something he never found. All he could think about was the well-bred horse and its elusive owner.

By the time evening arrived and the caravan came to a stop, Torren was feeling surly and sore. The nagging sense that something wasn't right had persisted all day, and he'd been unable to do anything to dispel it. He fumed in silence; and after taking one look at him, Larana handed him his dinner without a word and watched him with wide eyes from across the fire.

When the sounds of hooves came from the highway just about the time everyone was going to turn in, he wasn't surprised, though it did make his blood run cold. He'd purposely set their pit as close to the line of trees as possible just in case of this eventuality. Now, he packed what few of their possessions were out, his eyes locked in the direction of the road, as several of the guards got up to see who was traveling so late.

"Larana, let's go." He grabbed her wrist, not looking at her, all his attention focused on the highway. By the light of the fires, he could tell at least four riders had arrived, and they were speaking to the merchant. He couldn't see them clearly through the gloom and the people moving up to take a closer look, but he didn't need to. The glint of metal he'd seen as the riders came close was all he needed to know.

"Torren?"

A shot of fear tingled up his arm. He barely glanced at her as he led her back to the trees. Once there, he looked back to make sure no one was watching them. All eyes were currently on the caravan's visitors. With any luck, it'd be some time before either of them would be missed.

The darkness within the trees swallowed them whole. He led Larana deeper, trying to be as quiet as possible. He was starting to relax, thinking they might have gotten away before they could be discovered and cornered, when a dark figure suddenly stepped out from behind a large tree.

"Going somewhere?"

Torren's hand fell to his sword hilt as he yanked Larana behind him with the other. "Get out of our way."

Where there was one, there could be others. They'd planned for them to try to escape. He slowly drew his sword.

"We have no quarrel with you." The owner of the gruff voice shifted forward. "We only want the girl."

Torren tensed, feeling time slipping through his fingers but reluctant to attack just yet.

"We're willing to pay for her," the man added. "Name your price."

He felt Larana press up against him, her body shaking.

"I won't tell you again-get out of my way." He stepped forward, his sword held at the ready.

"Have it your way." The soft sound of metal rubbing

leather whispered as the man pulled out his own blade.

Still several paces from his target, he crouched, ready to go to battle, when Larana suddenly gasped out behind him,

"Torren!"

He glanced over his shoulder at the panic in her voice, and that's when his opponent chose to pounce. Torren caught the movement from the corner of his eye and, at the last moment, was able to block the deadly arc of his opponent's blade as it swung down for him. Behind him, he glimpsed another form trying to take Larana captive as she kicked and clawed.

His reach was shorter than his opponent's so he dropped back, acting as if he were unsure if he could best him. Taking his action as fear rather than the calculated move it was, the man lunged. Dodging, Torren found an opening as he overextended himself, and thrust his easier-to-wield blade under the other's guard. Like a snake striking, he slipped it just beneath the line of his opponent's leather chest guard.

There was a moment of resistance, and then the blade entered smoothly, warm blood shooting out over its length.

Torren drew his arm back, backing out of range, catching the shock on the other's face before the man doubled over in pain and collapsed onto the dark, leaf-littered ground.

Swinging around, he spotted Larana as she bit her attacker's hand and slipped out of his grip amidst heavy cursing. She'd almost made it out of reach when the man lunged for her and grabbed the back of her dress. There was a ripping sound as she was yanked back onto the ground with a thump.

Her attacker stood over her, fist raised; Torren slammed the hilt of his sword right into his face. The man's nose crumpled in a spray of blood, his head rocked back by the force of the blow. He fell like a sack of grain to the ground and didn't rise again.

Torren grimaced as he slipped his sword back into its sheath uncleaned before bending down and helping Larana to her feet. "Are you all right?"

She nodded, her breathing fast, her ripped dress half-falling from her shoulders. He took her hand and found it cold, confusion and fear flashing from the touch. He wiped a stray drop of blood from her cheek as he looked into her face, trying to confirm she was all right. Her frightened eyes sought his and seemed to calm down right away.

"We have to get out of here. Are you up for it?"

She nodded. "Yes."

Leading her away, skirting as far as possible both of the bodies, Torren penetrated deeper into the darkness, impelled by thoughts of pursuit. The two of them ran through the trees as fast as they dared, both keeping their senses alert for other unwanted strangers.

The clouds that had caused the day to be so pleasant thickened and darkened; and soon a thin mist that shortly turned into a curtain of water came down on them.

Though soaked to the skin, Torren was very grateful for the rain. If this group possessed as good a tracker as the last, rain was their best hope of losing them. It should make the tracker's job almost impossible. He felt himself grinning at the prospect. At least one thing was going their way.

A blurred amount of time later, Larana stumbled and fell with a cry behind him. He stopped, turning around to wait for her, this not being the first time.

Larana crawled shakily to her feet but then almost immediately collapsed again. He rushed to her side, worried for her.

"Are you all right?"

She didn't look at him, the rain dripping off her as she struggled once more to rise. This time he was able to catch her as her legs gave out.

"I'm sorry." She looked up at him, her eyes dark, her expression miserable. Exhaustion oozed from every part of her.

He frowned, realizing she'd valiantly kept up with him without ever saying a word.

"Let's rest for a minute. Then we can look for somewhere to hole up for the night."

She nodded, her body going limp as he set her down gently onto the muddy ground. He sat next to her, not far from her own level of exhaustion, the rain streaming off his face and hair. Without thinking, he brought her close, trying to shelter her with his body.

Larana drew a sharp breath, stiffening for a moment, but then gave in. In only moments, she had fallen asleep against him. Torren surprised himself by realizing he didn't really mind.

After a short while, he maneuvered her into his arms and carefully stood up. She didn't stir. Unsure as to exactly where he was going, he ventured on, trying to find them at least a partial refuge out of the rain. The best he was able to do in the darkness was a thick stand of bushes beneath a small outcropping. Doubting he'd be able to find anything better, he set Larana down, half-sitting. He unrolled a mostly soaked blanket and, after sitting down, covered both of them with it. He instantly fell asleep.

Movement beside him woke him up hours later. Torren opened his eyes to a much brighter sky and no rain. Larana was still sitting beside him, sleepily rubbing at her eyes.

He got up, glad to see the rain gone, and stretched his sore, tired muscles. The scent of wet clothes and drying leaves clung heavily to the air around them.

"I'll be right back."

Without waiting for an answer, he stepped through the bushes and out of their small shelter. Gazing up at what he could see of the sky, he made it out to be about mid-morning. He surveyed the area and found nothing out of the ordinary. Only then did he go behind a tree to relieve himself.

When he returned to the outcropping, he found Larana going through his pack, pulling out a change of clothes. When she dug deeper for some food, part of her back turned in his direction.

Where her dress was torn, a flap of fabric draped down, leaving about a quarter or more of her back exposed. Her white skin looked even more so in the light of the sun, except for two identical marks sitting to either side of her shoulder blades. They were both about a handspan in length.

Torren had never thought to see them in his lifetime.

A choking sound issued from his throat as he stood there, staring in total disbelief. Larana heard him and turned around, her face darkening with apprehension.

"Torren?"

How? How was this possible? After all this time, all these years! He shook his head slowly, his numbed mind fighting against what he'd seen. Such a thing shouldn't have been possible.

"Torren?" Larana, openly worried now, took an unsure step toward him.

He took a step back, but that was as far as his legs would take him. They crumpled beneath him and dumped him onto the damp ground.

She rushed forward, reaching out to help him. "Are you all right?"

"Don't touch me!"

Larana stopped, dumbfounded, and pulled back at the unexpected vehemence in his voice. "What's wrong? Torren, what's happened?"

He turned away from the sight of her, putting his face in his hands. He felt hot, cold. What kind of farce was this? He'd *found* her? He leaned abruptly forward, nausea racking through him. His chest hurt. His hands dug into the damp dirt, pulling up the soft matter as he curled them into fists.

How many times had he fantasized about this moment? In how many of them had he found her and killed her, found her and loved her, found her and killed himself. And now—now to find she'd been with him all this time, with him totally ignorant of whom she really was.

The irony of it burned him. If not for what had happened the night before, he might have never known. A rough bark of a laugh escaped him. He would have never known.

Almost against his will, his head turned so he could look over his shoulder. There she stood, looking concerned and innocent, the catalyst of his current life, the bane of his existence. A maelstrom of emotions swept through him, threatening to drive him mad. Joy, anger, love, hate, triumph, loss—how could he feel these things all at once?

"Torren?" Her arms about herself, looking hurt and confused, Larana took a tentative step toward him.

"Stay back!" He glared at her, his body shaking, a war he'd not expected waging inside him.

Who would have ever given the scrawny, gangly girl before him a second look? Only the markings on her back revealed who she really was. Surely, those who'd raised her had seen them, had known. How could they not? How could *she* not? Had she been playing with him all this time?

It took all he had not to confront her. She didn't know. He could tell from her voice, the expression on her face. She was as much a victim as he was. Part of him was galled at the thought while another part latched onto it as if it were a rope for a drowning man.

But to have found her! Him, of all people—it would be laughable if it didn't hurt so damn much.

Torren looked away from her, from her innocence, from her concern, from her unknowing culpability, and let out a long, shuddering breath. By pure force of will, he made himself rise to his feet while everything swayed around him.

"We need to go. They'll be looking for us."

He could feel her eyes boring into him, wanting—needing—some kind of explanation. Something—anything—to explain his strange behavior, but it was more than he could bring himself to give.

Without looking at her, he turned to retrieve his pack, taking care not to get too close. He put it on, though it almost unbalanced him, and started off without another word. He heard Larana follow him. It made different parts of him cry with joy and despair.

This scrawny, clumsy girl—even her coloring was wrong. They weren't normal. The Chosen were fair-haired and blue- or green-eyed. Still, he'd heard there had been others who'd not conformed to the norm.

Those who'd raised her had to have known. Even grubs were aware of the general history of the Flyers, despite the more horrific tales they'd devised about them on their own. So, why had they kept her? Had they wanted a child so badly?

Her ignorance smacked of duplicity on their part—their

eagerness for her to hide her strange powers, her almost total isolation from others. Did they just find her, or had there been more to it than that? Larana's story of her last night at home whispered of collusion. But why? Why?

When he finally called for a break, Torren couldn't even bring himself to glance at her. He took out food and water and set them out where she could reach them yet stayed well away. He stared at their meager supplies and scowled. He should have planned for this eventuality better. But then, he'd been a fool for quite some time, hadn't he?

"Torren?"

He stiffened at the sound of her soft voice.

"We'd better get moving." He pushed away from the tree he'd been sitting against, getting ready to go. He ignored the fact he'd not given either of them time to eat or rest and set off, leaving her to scurry up behind him.

He kept traveling throughout the day and didn't stop until it was almost fully dark. During what breaks they took in between, Larana didn't attempt to disturb the silence between them. It was just as well. He had nothing he could say.

Camp was made in a copse of close-growing trees that would keep them out of sight. He handed out Larana's portion of their meager dinner, stealing only the barest of glances in her direction. Her legs, dress and feet were mud-splattered, damp leaves sticking to her here and there. She still carried her change of clothes, though she'd at least put her vest on, removing her back from view. Her hands shook as she reached for her food, her face pale and her eyes red, as if she'd been crying. He made sure not to glance her way again.

Setting out a blanket for her and taking the other, he sat up against one of the old trees and wrapped the cover about him. He heard her lie down, and was about to drift off to sleep when her whisper hauntingly to him from the dark.

"Please, Torren, tell me what's wrong." There was a note of supplication in her voice that made him wince.

"Nothing," he said, a little more harshly than he'd intended. How could he explain this to her? "Nothing's wrong. Go to sleep."

"That's a lie!" Her anger and torment stunned him. "Is there something wrong with me? Did I do something? I don't understand!"

It was more than Torren, in his current state, could bear. He straightened where he sat, bunched fists at his side. "And why don't you? Aren't you supposed to have the knowledge of the gods? Don't your strange powers tell you everything you need to know? Isn't that how you found me in the first place?" He was screaming at her, screaming at *her*. But he couldn't stop. He couldn't—all of this had been bottled up inside him for too long. "Isn't this all just some sick joke to you?"

A tortured sob bridged the space between them. "I... I..."

"I should have just let them have you." He was horrified at the words leaving his mouth.

In the darkness, there was a flurry of movement and sounds of crying as Larana leaped up ran. He stood, trying to determine which way she'd gone. He should go after her. Apologize. But a part of him declared she deserved no better. Let her sleep amidst the trees if that's what she wanted. She had nowhere else to go. She'd be back soon enough.

# CHAPTER 1

TORREN BARELY SLEPT, HIS SENSES ON CONSTANT ALERT for Larana's return. As the sky finally lightened, her empty blanket stared at him accusingly. He glared back at it, almost asking it how it had expected him to find her in the darkness. He would have become completely lost. As lost as she probably was.

He sat up straight, a jerk of panic and joy coursing through him at the thought. She was a farm girl, sheltered, never been far beyond her home. How much woodlore would her keepers have taught her? Had she changed her mind and tried to come back only to find she had no idea how? His chest grew tight.

"Larana?"

He got no response.

"*Larana!*"

She was gone. She'd finally done for him what he'd been trying to arrange for days. But she had no supplies, no idea where she was, no money. Yes, but she was gone—who she was and what she stood for were no longer his concern. Shame flooded through him at the gleeful notion. She was innocent, no matter who she was. She would die out there alone. What would it make him if, after all he'd lived through, he allowed her to perish this way? If he let her die because of his anger, he would be less than nothing. His father would curse him from the grave.

Moving stiffly, he picked up her blanket and his own, rolled them and attached them to his pack before crouching and carefully examining the ground. There, close to a large oak, he found an imprint of a small shoe leading away from the camp.

Anger mixed with dread, all coated with worry. Guilt wore at him as he set out to track her down. His thoughts ran over the events of the night before and informed him yet again he was the cause of this. Larana possessed no idea of what was wrong. She must think him a madman. But how could he explain it to her? How?

He'd been searching for her for so long. Not always consciously, not always willingly, but he admitted not to himself that it was true. Not once had he actually believed he'd find her. Yet now he had and, through his own stupidity, lost her again. His past wasn't her fault, no matter how easy it was to blame her for everything.

What, though, would he do with her?

Larana's path was erratic. She'd stumbled often, and he picked up speed as more and more clues of her passing appeared. The path twisted and turned, as if she'd been trying to get back to him but couldn't find her way. Had she called out for him? Had she thought him near but tormenting her by not responding?

The sun stood high in the sky when he finally tracked her down. He found her curled up between the roots of a large tree, a squirrel sitting on top of her as if claiming her for its own.

He rushed forward, startling the squirrel, who scurried up the tree and out of sight.

"Larana!"

He got no reaction as he knelt beside her. Shame and exultation clashed inside him when she didn't react. He quickly noticed she was paler than the day before. Her eyes darted beneath closed lids. He tried to push everything inside him away and reached to touch her, even as part of him insisted he not do so.

Her skin was damp, and felt hot to the touch.

"Larana."

His hand tingled as her eyes flickered open. Her blue eyes were bright, too bright. "T–Torren?"

"Yes, it's me," he said gently. "You'll be all right now."

Tears welled in her eyes, her cheeks suffusing with

color in two round patches. "Please...Please don't hate me."

Torren jerked his hand away, having forgotten about her power. He stared at her, this poor, pitiful girl with no one to depend on but an embittered mercenary. Abruptly, the battle inside him came to a stalemate. With a chill, he accepted she was as much, if not more, of a victim than he was. It would have to do for now.

"I...don't hate you. This has nothing to do with you. The problem's mine. It's not your fault."

His words and his attempt at a soothing tone must have told her what she wanted to hear, for Larana closed her eyes and her flushed face began to clear. He shook his head, wondering how, after all this, she could trust and believe in him so easily.

Removing the blankets, he wrapped her in them and gently moved her into a patch of sun. Digging into his pack while keeping an eye on her, he brought out a wrapped bundle of dried herbs. He sifted through the small stash, mentally kicking himself for not having more. Fighting wasn't the only thing Sal had taught him. He just hoped the lasa leaves were still fresh enough for what he needed them to do.

When freshly picked, lasa leaves were potent enough that all you needed to do was chew them to get their medicine coursing through your system. Once they'd dried, though, it was normally necessary to boil them to bring the medicine out. Between the heavy rains and the fact they were being hunted, building a fire was something he couldn't afford to do. Instead, he crushed the leaves into a cup, and using the end of his dagger ground them as close to powder as he could. He filled the cup with the last of their water, hoping that at least it would do more good than harm.

Larana moaned softly as he shifted her weight to his lap so she was partially sitting up. Gently, he parted her lips and trickled the concoction into her mouth. He felt the muscles on his face relax a little as she automatically swallowed. Maybe the fever wasn't high enough to be beyond the leaves' power yet.

Putting everything away, he picked her up and staggered off. The clouds were starting to look ominous, so he needed to find some kind of shelter lest all his ministrations be for naught.

Arms aching, he finally found an old dead tree with a hole on the side. Setting the girl down out of the way, he knelt and hacked at the opening with his knife to make it bigger. The sky continued to darken. Torren kept gouging out the soft rotted wood, feeling time slip away.

The first drops were just falling when he stopped, his arms shaking from exhaustion. He fell to the ground, eyes closed. A large drop landed on his nose, the air drenched with threatening moisture. He wallowed in the cool sensation on his heated face but groaned a moment later as he forced himself to get up. Stumbling over to pick up Larana, he scrunched into the broadened hole, sweeping away what termites still fell from the tree's disturbed carcass. He dragged her wrapped body in after him and pulled her against his chest. He then dragged in his pack to block the opening as much as possible.

He'd barely gotten situated when the rain started coming down in earnest. It poured for the rest of that day and into the night.

# CHAPTER 11

Torren's eyes flickered open as the comfortable warmth covering him moved away.

The first thing he saw were Larana's sky-blue eyes staring at him intently from less than a hand's-width away. The girl looked quickly away, appearing as shocked as he felt at being in such close quarters. A healthier blush than the one she'd had the day before tinted her cheeks.

She shifted nervously, causing him to grunt in discomfort.

"Go that way." Trying to point, he kicked out the pack covering the entrance, giving her a way out. She bolted, stepping on him in the process, the two blankets around her going every which way.

He followed more sedately, his legs screaming as blood flow returned to them. He was forced to sit down outside, grimacing, until the painful tingling subsided.

When he was finally able to stand without pain, he found Larana perched several lengths from him, watching him closely while trying to pretend she was doing no such thing.

He was glad to find the torrent of clashing emotions that had so overwhelmed him the day before quieter. At least now he would have some room to think.

"Are you hungry? You haven't eaten anything for at least a day."

Like a startled animal, she twitched at the sound of his voice, as if not sure what to expect. She stared at the ground a moment then lifted her eyes, a determined look on her face.

"I want to know why."

"Why what? Why you should be hungry?" He reached for his pack, feeling awkward at the evasion, knowing exactly what she wanted to know.

She shook her head. "No. Why...Why you looked at me the way you did. Why you think you're a joke to me. I wouldn't even be alive if not for you."

"Here." He held out a portion to her, and hesitantly, she came and took it from him. She didn't eat, still waiting for his response.

He wasn't quite ready for that yet. "We'll be heading back to Caeldanage. With the rains, I'm hoping most of our trail will have been erased and they won't figure out where we're going until we're safe."

Larana nodded, still looking expectantly at him.

He sighed and looked away, finding her continued blind trust disturbing. She'd thought he hated her less than a day ago, and in some ways she'd been right. But now, though he was acting more demanding than before, she still didn't question his decisions. Of course, it might not be trust at all, only the fact there were no other options open to her. He sighed again.

"Do you know about the birthmarks on your back?"

Larana nodded slowly, appearing confused. When He only stared at her and said nothing else, she finally managed to speak.

"Yes, I know about them. I don't really know what they look like, but..."

He nodded. Wondering why he was doing this, he picked up a stick from the ground and drew two equal but opposite images on the muddied ground. "This is what they look like."

Larana stepped tentatively forward to take a look. "They seem so big." Eyes wide, she glanced up at him. "Would...Would you show me? On my back?"

He felt himself turn cold and then warm again. Reminding himself she knew nothing about any of this, he gave in to her request. "Turn around."

She did as he bade her, removing her mud-stained vest

and revealing her torn dress. Torren swallowed hard as the twin birthmarks came into view. With a shaking hand he was grateful she couldn't see, he reached out with a finger and slowly traced their patterns for her.

Once he was done, Larana spun around and stared at him, a strange look on her face. He could make nothing of it, was only barely aware of it as he stared at his own hand, thinking of what he'd just touched, and tried to pull his thoughts together.

"Are they...? Do they mean something?" Her eyes searched his face, as if she already held an inkling of what was coming.

He nodded, half-turning away. "Do you remember the things I told you about El and the Chosen?"

"Yes."

Torren glued his gaze to the ground. "Well, there was another gift El gave his people I didn't tell you about." He paused to glance in her direction. "He gave them his Vassal."

"A vassal? I don't understand."

He fought for the right words, part of him insisting he must tell her while another wished he didn't have to try at all. "El's Vassal—he or she is the Chosen's ruler, their spiritual guide. There has always been one, for when one dies another is born to replace him...or her." His voice grew quiet. "The Vassal has gifts the Chosen don't, but also has no wings. Instead, El's mark is on him so there won't be any doubts of who the Vassal is. These marks reside in the same place where his wings would have attached were he just a Chosen."

"Oh."

"Marks the same as yours."

Larana sank to the ground. "But...how?"

"El's Vassal is the Chosen's most sacred possession. He or she is their link to El. The Vassal proves El is with them and knows of their travails." His throat felt dry. "After the sudden death almost fifteen years ago of the last Vassal, a new one was born—a baby girl. She'd not been at the capital long before she disappeared, stolen by the grubs—or so it is said."

He turned to look at her, his eyes locking with hers. "It

would seem the reason those men are so eager to get hold of you is because they believe you to be the long-lost Vassal of the Chosen."

Larana looked away, her face a mask of uncertainty and disbelief. "Do you...Do you believe I am this person?"

He forced himself not to turn away, though he wanted nothing more. "Yes."

She stared at her dirty hands on her lap. "But..."

"Yes."

She glanced up at the force in the one word and slowly shook her head.

"It can't be!" Some of the intensity she'd displayed when he'd returned for her at the inn flared then left again. "It's just a fluke. My birthmarks only resemble the Vassal's."

Her eyes begged him to agree.

Torren only stood, his face set. "It's you. There's no mistake." He knew this to the bottom of his soul. No matter how it made him feel. "And that's why we're going back to Caeldanage. The Chosen are there, and they will want you back."

Larana hugged her legs to her chest, looking suddenly small and alone. "You...You didn't know. Not until you saw my back. And then...And then you...Why? Is there something more to being a Vassal? Something evil?"

He turned his back to her and retrieved his pack. "No. There's nothing evil about the Vassal. I told you before—the problem is mine. It has nothing to do with you." He glanced back at her and saw the open confusion on her face. "You'd better eat now. We have a long way to travel today."

Nodding, she ate hungrily. Torren amazed himself by almost smiling. Her ravenous appetite was a definite sign she was doing better.

He set a slow, easy pace, making sure to stop for rest often. He made no conversation, and Larana seemed to be too absorbed by her own thoughts to mind. He thought it just as well. The Chosen on the floating island, or those at the embassy, would tell her all she needed to know.

Around mid-afternoon, they reached a stream overflowing its banks because of the large amount of rain. Too small to hold fish, it was nevertheless fresh and good for bathing.

"We'll make camp here. And since we still have a few hours of light, I'll go scavenge us something for dinner while you clean up." He glanced over at her. "If you're of a mind, you might clean our clothes as well."

"Leave it to me," Larana responded eagerly. She took his pack when he offered it, her face brightening at having something to do.

"I won't be gone too long," he told her then turned to follow the stream. He'd already walked out of sight when it occurred to him he'd just asked the Vassal of El to do his laundry.

His heart skipped a beat even as he half-smiled. Though he was sure to the core of his being Larana was the one, it seemed a part of him still thought of her only as the young farmer's daughter. It only served to reinforce the fact he had no right to blame her for anything.

It wasn't long before he found what he was looking for—fresh animal tracks leading to the water. Catching game was out of the question; it would be too time-consuming, and it wasn't safe to build a fire. But by back-tracking the trail he could find where the animals had dug for roots or pilfered nuts, and some of this fare would supplement their dwindling supplies nicely.

Once he had collected enough to satisfy him, he returned to the stream and bathed. As the sun moved to hide behind the horizon, he reached their camp with an armful of roots and a few berries. When Larana spotted him, she jumped to her feet, the relief obvious on her face.

Their laundered clothes hung neatly from several low-hanging branches, drying. She was clean as well, her hair washed and retied in its usual thick braid. Her face and arms were free of mud, and she looked almost pretty.

He'd found her; he'd been the one. Or, to be more precise, he amended a moment later, she'd found him.

"These should be safe to eat," he said, showing her his bounty. "I've already washed them."

She gave him a smile of appreciation for his efforts, and he felt strangely grateful to her in return. They quickly divided the spoils and, after supplementing them with some dried fish from his pack, ate in silence as the light in the forest slowly bled away.

"Torren?" Larana called, her voice meek.

"Yes?"

"What will happen to me?"

He pulled the blankets from where she'd hung them to dry, mulling her question as he felt a frown mar his brow. "Nothing will happen to you. You'll just be reunited with your people. You'll be protected, pampered. You'll get to travel on the Twenty Islands, happy and safe. You'll help the Chosen lead the best lives they can."

He glanced in her direction; and though her face was now dimmed by shadows, he could see she was pondering his words.

"I can't imagine it." She shook her head. "I'm a farm girl. I've never been anywhere, seen any wonders. I'm not wise. I'm clumsy. I don't even know anything about the Chosen"

"You'll do fine," he told her. "There'll be people there to help you through it all. And there'll be one other bright side in it for you."

"What?"

He felt a pang in his chest as he gave her a gift she could not have considered. "You'll get to meet your parents, your brothers, your sisters—your family. Your real family. You'll once more have a place to belong."

"Oh." Her features cascaded through a number of emotions as she realized the ramifications of what he had just said. "Oh!"

"Goodnight." He tossed her a blanket and lay down on his own, filled with bittersweet amusement.

Over the next several days, Torren guided them south. They avoided roads or towns except for a couple of times when he left Larana hidden to buy needed provisions. They continued not to light any fires at night; but since the evenings weren't overly cold, this gave them little trouble.

Larana became a whirlwind of both excitement and worry whenever the floating island appeared as they came closer to Caeldanage. She asked many questions about the Flyers, but he told her little. It didn't stop her from asking more, though.

As the tree line dropped off, the city became a beacon in the distance. Torren stayed in the fields, seeking what shelter they could find amidst the maturing crops.

Eventually, they came abreast of the city's towering walls, the island floating arrogantly above it. As they approached, he didn't hide anymore, only too aware of the eyes watching from atop the walls and what they might make of such odd behavior. He'd managed to buy some cloaks to make up for the ones they'd lost, which helped give them some anonymity.

Following the wall, they headed for the highway. As they drew near to it, he stopped, and they sat up against the wall.

"We'll wait here till it's almost dark and the gates are about to close. We should be able to lose ourselves in the dark once we get inside if the men looking for you spot us."

Larana stared at him, looking worried. "You think some of them are still here?"

He shrugged. "There's no way to know, but they've gone through an awful lot of trouble so far. Luckily, they don't know exactly what we look like. And our coloring is enough alike most people will assume we're related and not give us much thought."

She nodded, trying to appear calm, the rapid tapping of her fingers on her knees betraying her true feelings.

"Don't worry, in a few hours time, you'll be safe."

A couple of hours later, Torren stiffened as the floating island's shadow drifted over them. He would deliver the Vassal to them, but he was doing it for her, not for them. He could have definitely done without coming back here again.

As the sun sank in the horizon, he stood and signaled Larana to do the same. Making sure their hoods were drawn up far enough to conceal their faces, they started off toward the highway and the gate.

"Hunch over a little and cough as we go in. If you look ill, it'll explain why your hood is up even though it's not cold."

Larana nodded, falling slightly behind, and did as he asked. He stopped at the gate as the guards were getting set to shut it and let her catch up. He put his arm around her shoulders as if helping to support her as they made their way inside.

No one bothered them as they entered, in fact, barely spared them a glance. He kept his eyes focused straight ahead so he wouldn't attract attention and led Larana off in the direction of the apothecary's quarter. As soon as they were out of sight of the gate, he pulled her into a dark doorway and waited, his hand on the hilt of his concealed sword.

Larana stuck close to him, and inhaled sharply as a man appeared from the direction they'd come, glancing about as if looking for someone. Torren tensed, ready for trouble, though he expected none. The man, not spotting them in their dark niche, ran down the street in the direction they were likely to have gone, his soft-soled boots making little noise.

As soon as he vanished, Torren took Larana's hand in his and got a jolt of her trepidation.

Sticking to the shadows, he took a different way into the interior of the city. The floating island's towering presence and the shadows it cast made the way quite easy, almost as if it knew the Vassal was on her way back home.

Most shops were dark, only the windows in the upper-story living areas showing any light. Taverns and inns were lively, snatches of conversation, music and laughter spilling out into the streets. Several of the temples dispersed throughout the city were lit brightly from within, as if shunning the dark presence of El's people above. Prayers drifted in the wind from the Temple of Valem.

Torren felt a faint shiver scurry down his back and wondered if those inside were wishing ill to the Chosen—or, more specifically, to the Vassal. Though Valem's people had never done anything directly to the Chosen, aside from Valem's own actions against El, they didn't look upon each other kindly.

Which part of the empire one was in also determined how hostile the relationship between the Flyers and

Valem's followers was. He had heard of several instances where religious influence was used on governing bodies to try to get them to curtail associations with the Chosen, even when this wouldn't be in their best interest. Fear was a great motivator.

To everyone's good fortune, a god of fear and death wasn't usually a popular figure, so his followers were few. Torren was sure most of the stories about what they did behind their dark walls and the slashed and burned effigies of the Chosen they were supposed to indulge in were no more than unsubstantiated rumors.

Still, the Chosen's supremacy in the air and seeming immunity from the laws and restrictions of Landers and their countries hadn't made them popular in many eyes. If not for the fact the Landers had yet to find an easier, more cost-effective way to transport large or heavy goods to other places, things might have been more strained.

For anyone but the gods, magic was hard to gather and even harder to use. The possibility of replicating El's creation of the islands and the flying ships was beyond imagining. But magic was not the only way. Down in the southern states he'd heard of attempts to replicate the wings of birds through science. Torren was sure nothing they came up with, however, would ever compare to the feeling of having the real thing.

Aside from their progress being slow and the need to hide and wait several times as people or guards crossed their path, they encountered no problems in their travels through the streets. Their circuitous path finally took them into the richer part of the city. Shopfronts were less evident, and the homes flaunted gates and walled-off gardens. Street lighting grew more common, making it harder for them to stay out of sight.

Torren slowed as they climbed a small hill toward a multistoried residence. The entire estate was walled; the entrance sporting a metal gate as well as two thick, reinforced doors, almost like a fortress. A large gilded symbol of a pair of wings adorned the doors, leaving no doubt as to who resided there.

During his many forays through the familiar city, he had only come this way once, yet the embassy's location was burned forever into his memory. Never in all his years had he believed he'd willingly come to these doors again.

Glancing back at Larana and seeing in her face the same nervousness he felt, he reached with a suddenly moist palm through the bars of the gate to pull on the cord of the summoning bell. He flinched, the clanging of the bell echoing loudly in the empty street. Larana stepped in close, scanning their surroundings as if expecting men to leap at them from the shadows.

After what felt like an interminable amount of time, part of which he spent considering whether he should ring the bell again, heavy footsteps approached from the other side of the gate. Torren placed himself in front of the cloaked girl, wanting to keep her out of sight. His breathing sounded heavy in his ears, and his heart pounded hard inside his chest as he heard unseen bolts being thrown. Moments later, the left door opened a crack, letting a shaft of light fall over them.

He blinked, not bothering to cover his eyes, for the first time in more than half his lifetime standing face-to-face with a Chosen.

"State your business." The lightly accented request came from a man close to Torren's height. He wore a plumed silver helm with a nose guard that hid most of his face. He wore a chest piece made to look more like a bare-chested man than armor, and protective pieces covered his thighs and legs. He stood in profile to the door, his wings swept back, the hilt of his short sword well in evidence. Torren knew from the color of the guard's armor he held rank.

His tongue lay thickly in his mouth, but he pushed himself to speak even as the guard's blue eyes studied him up and down.

"I must speak with the ambassador."

"It's after hours. The embassy is closed," stated the guard impatiently. "Come back tomorrow."

Torren forced himself to try again, though a part of him wanted desperately to latch on to the offered excuse.

"I realize that, but circumstances warrant I see him now."

The guard eyed him with suspicion. "And what dealings of a Lander might be so pressing?" He sounded in-
creasingly annoyed.

Torren used the only thing he had. Anything less was
likely to get the door slammed in his face.

"I have information on the Vassal."

# CHAPTER 12

THE GUARD FROZE; HIS EYES WIDENED AS HE MOVED his attention from Torren to the half-seen cloaked form behind him.

"Wait."

He closed the door, and they heard the bar drop back into place.

Once his footsteps receded, Larana whispered, "Do you think they'll see us?"

"Of course." In reality, he wasn't so sure. As the minutes ticked by, he felt more and more exposed in the open, too-quiet street.

If those pursuing them possessed the faintest inkling he might know who Larana was, they could have the area around the embassy watched and grab them off the street, leaving the Flyers none the wiser. They'd be fools not to at least have the approach to the place kept under observation. He and Larana might have been able to sneak past under cover of darkness, but if they were forced to come back tomorrow there was no way they would be missed.

"To—Torren, tell me more about El, about the Chosen." Nervous fear coated her every word. He knew only too well how she felt.

In soft whispers, he tried desperately to distract her as it appeared ever more likely the guard wouldn't come to let them in.

"When El returned to the First Mother and those of his kind, he gave the Chosen all the gifts I mentioned before except for the Vassal. Some of the other gods, upon learning what El had gone through and done, also decided to spend time with the humans, but as their betters. For, while El gave up his powers to become human, the others didn't want to be that vulnerable. They had no desire to feel the fear or helplessness so alien to their natures.

"Chaos ensued. The gods didn't see the humans as be-ings but as playthings. By keeping some of their powers while at the same time becoming more human, the les-sons El learned were harder to come by. Instead, the gods began a great competition, each trying to acquire more followers than the others. They manipulated people, They caused great wars between nations, They made their sub-jects build huge monuments in Their names so they could be worshipped and praised as creators, though they knew very well the First Mother created all."

He dredged up the story, which he'd thought long for-gotten it. It frightened him a little, though he was loath to admit it to himself, how easily it all came back.

"The Chosen were safe, looking down from their is-lands, not understanding the foolishness going on below. Some attempted to help a few of those suffering, but after being caught once or twice in violence propagated by jeal-ousy or mistrust, they kept to themselves.

"Though it had been long enough for many of the Cho-sen to forget or deny their origins, El had not forgotten. He knew those humans dying because of the pettiness of his brothers and sisters were the same as the humans he'd chosen for his own. And while he could do nothing to stop them on his own, he knew who could. El begged the First Mother to intervene.

"The First Mother asked her children to come home to hear El's words so there would be understanding among them, but some of them refused to come. When the Mother entreated them again to return, Valem rejected her, saying he didn't need her and had even less need for El. Valem and several others had grown to enjoy the pleasures of the flesh, so they had no intention of revert-ing back to what they'd been.

"Since the time of the Grand Creation, the First Mother had never been rejected by one of her own. It made her angry. And her anger was felt in the heavens, in the ground, from the lowest to the highest, as all she'd made shook with her fury. The next time she didn't ask, didn't entreat, but forced her children to come home, changed them back to what they'd been, giving them no choice. No longer would they be allowed in the world, no direct contact would they be permitted with the humans. The gods were her children, and they could not do without her yet, nor would she allow them to disavow her or corrupt her creations.

"But at this El protested, for he loved his Chosen. He loved all the First Mother had made and didn't want to be forever denied them. So, he opened himself to her, so she would see what they meant to him, how the Chosen meant as much to him as her children meant to her. And so overwhelming was what he showed her, the First Mother relented a little. Instead of totally depriving the gods of the world, She decreed they could still see it, but they would no longer go there in a guise of flesh or interfere with those who lived there directly.

"It was then El granted the Chosen his most heartfelt gift—his Vassal, a link through which he could indirectly guide his people and also give them a living symbol of his love."

Torren tried to look at Larana's face, but the shadows inside her cowl were too deep. He grew silent then, turning once more to the closed gates.

Two of the moons had risen a long way into the sky before footsteps once more rang on the other side of the doors. Torren tensed, not sure of what would happen next.

The wooden doors swung wide. Four guards stood there, one in silver-colored armor, the other three in bronze. The three held spears at the ready as the leader stepped forward with a light to open the metal gate.

"The ambassador will see you," he told them quietly.

Torren nodded, putting his arm protectively about Larana's shoulders.

"First, however, you will remove all your weapons and allow yourselves to be searched."

Torren's narrowed his eyes, not liking the unusual precautions. Had someone gotten to them before they'd arrived? Flyers weren't known for being cautious.

"Why is this necessary?"

Sharp blue eyes met his.

"Circumstances warrant it." The commander's wings jerked once. "You asked to come in. If you don't agree with the conditions, you don't have to stay. It's your choice."

Torren exhaled slowly, telling himself this man was only doing his job. He removed his pack, then his sword, and finally the knife in his boot. He passed the two weapons to the guard through the metal gate but set the pack on the ground, as it was too big to fit through.

Once this was done, the commander removed a key and opened the gate. He grabbed the pack and handed it to his fellows then opened the gate wide and gestured for them to enter. Surrounded by the guards, Torren stood while the commander patted him down for concealed weapons.

Finding nothing, he motioned for Torren to step back and for Larana to come forward. Her features still hidden in the cowl of her cloak, she stepped through the gate.

He patted her down without removing her cloak, and as he did she sneaked a hand out and caressed the end of one of his white wings. The commander jumped back as if he'd been slapped, startling all of them. The other three men rushed forward even as Larana backpedaled.

"I'm...I'm sorry!"

Torren prepared to step between them.

"Stop! It's all right." The silver-armored guard held up his hand, his voice tinged with wonder. "She just surprised me, that's all."

The other three stared at him for a moment then backed down.

"Please, come this way." The leader's voice showed a deference it hadn't before.

One of the guards stayed behind to close the gate and the doors as well as appropriate Torren's possessions. Passing out of the deep entryway into the walled grounds, they were escorted down a winding walk to the columned porch of the opulent embassy. The gurgling of a fountain could be heard somewhere off to the right. The scent of flowers and freshly cut grass filled the air. Torren studied the two-storied building before them.

It was purported to be an amalgamation of Lander and Chosen construction styles, a gift from the emperor in an attempt to solidify relations with the Chosen. The columns were in the Flyer style, starting out with a broad base, narrowing down to half the width not far above, and then slowly widening again at the top. The portico and part of the roof were in the Lander style: thick, bracket-shaped tiles lapped over one another.

Larana stared up at the building, the cowl of her cloak falling back to reveal her curious face in the lamplight. He thought he heard an in-drawn breath from the silver-armored man.

The square doors were opened for them, and they were led inside. Their footsteps echoed as they followed a short hallway to a set of double doors on the right.

Two of the guards opened the doors wide, while the commander removed his helmet as he stepped in. Curling golden-blond hair glinted in the light; and it was a young man, possibly no more than twenty, who watched them as they entered.

The doors closed quietly behind them, the two guards who'd opened them remaining outside.

The room wasn't as well lit as the hallway, shadows falling across several plush chairs with the narrow backs so common to most Flyer furniture. Standing by a small lit brazier was a winged man wearing ruby-colored robes. He turned to face them as they moved farther into the reception room. His deep-green eyes in a lined face studied them as he sipped a cup of mulled wine.

"Uncle, I've brought them as requested."

The ambassador nodded, his eyes not straying from the strangers.

"It is my understanding you desired to see me?" His accent was less pronounced than the guard's.

Torren blinked, saying nothing, caught off-guard by the ambassador's bored tone.

"Well?" he prompted impatiently.

Torren bowed. "Yes, sir, such is the case."

The ambassador's brow rose at the gesture. He turned to face them fully for the first time. "And what has brought you to me at such an hour?"

Torren stared into the ambassador's eyes and trapped them with his own. "I've brought someone in need of your

112

protection, someone you've been seeking for a long time."

A look of incredulity flashed on the older man's face. "And just who might this be?"

Torren motioned Larana forward. She stared at the ground.

"Hmph, I see a typical Lander child. Why would we be seeking her?"

He said nothing, only nodded to Larana, who meekly removed her cloak. Without a word, she turned around and removed her vest. Torren stepped forward to sweep her thick braid out of the way and pulled down carefully on her collar to the right. The tip of one of her birthmarks came into view.

The ambassador's gasp filled the room as his goblet fell forgotten to the floor.

"Uncle?" The young man sprang forward, not sure of what had startled the older man. As he turned to look at what his uncle had seen, he stopped, his eyes growing wide. "By El's will!"

Torren released Larana's collar and stared at the two men. The girl turned around and watched them, too, her eyes filled with both excitement and fear.

"After all the impostors...how?" The ambassador took half a step forward and stopped. He shook his head, as if trying to bring himself out of a dream. "It will have to be fully verified. We can't afford to make a mistake, not with this."

The young guard stared at his uncle. "But if it's her..."

Two similar smiles flashed momentarily on their faces.

"Yes, Micca, we would all be vindicated. Things would be set right again after all this time." The ambassador glanced in Larana's direction, his eyes bright. "My name is Rux. How are you addressed, miss?"

There was a strange expression on his face as he asked this.

"La–Larana." She tried to give him a curtsy and only botched it a little.

Rux's expression darkened for a moment, though it was hard to tell if it was due to her clumsiness or something else.

"It is my privilege to meet you, Larana. I thank you for

113

coming. And if you wouldn't find it too impertinent of me, I'd like to request a favor."

She blinked, startled by his change in attitude. "Of course."

Torren stood back to lean against a wood-paneled wall in the shadows, knowing what would come next. Soon there would be no doubt in anyone's mind as to who she was.

"We need to inspect the marks on your back," Rux told her. "To do so and not embarrass you, I'd prefer to lend you one of our robes." He half-turned, his strong wings lifting apart to give her a view of his back. "The way they are made, once folded over the body they leave free the area around the wings. This would easily allow us to study your markings and nothing else."

The girl frowned, staring at his garments, then nodded. "Allright."

"Micca."

The young guard nodded, sending a friendly glance in Larana's direction. "I'll get it at once."

He almost ran to the double doors.

Torren said nothing in the intervening minutes. The ambassador seemed unaware of his presence, all of his attention focused on the girl. Larana fidgeted at the attention, her eyes roaming the room in curiosity.

Micca returned presently, half out of breath, carrying a folded garment the color of eggshells.

"It was the smallest I could find." He handed the bundle over to his uncle.

"Thank you." Rux turned to Larana. "My office is through that door." He tilted his head toward a single unassuming door on the right wall at the end of the room. "You can change there."

Larana threw a glance at Torren; and when he did nothing and said nothing, she took the robe.

"You might wish to show her how it's worn—she won't know how."

Both Micca and Rux glanced at him in astonishment at the unexpected interruption.

"Is that so?" Again the strange look flickered over the ambassador's face. Quickly hidden shock reflected on the

114

younger man's. "We shall have to remedy that."

Signaling to Micca to help him, Rux unfolded the robe. The cut was distinctly unusual. The main body was a single large piece, but at the top was a long strip of cloth with a hole for the head, while from the sides of the main body flared four strips at an angle, two from each side.

Using Micca to demonstrate, Rux placed the hole over the former's head. The rest of the strip was allowed to hang down the back while the main body hung on the front. Next, Rux took the left side of the garment and draped one of the lengths over the wing on that side while the other went beneath it. The lower was then draped over the opposite shoulder and the higher beneath the other wing.

The same process was repeated on the right side, the cloth overlapping. In the front, the lengths were wrapped yet again and then tucked into one another.

Going slowly, Rux repeated the pattern in reverse. He then handed Larana the robe. She stared it, looking unsure, then gave Torren a final sideways glance before heading off toward the office.

As soon as she was gone, Rux stepped over to a chair and fell into it. Micca stared the way Larana had gone, an irrepressible smile fighting to fill his face.

"The Vassal."

"We will find out soon enough," Rux said quietly. "There have been others before her."

"Yes, but none have ever known about the marks!" Micca added eagerly.

Rux slowly shook his head. "But she seems utterly ignorant..."

"It's her, it's got to be!" Micca stared at the closed door.

"It appears there was more validity to the rumors Valerian spoke of to the council than I at first believed."

Torren frowned. They'd known she was here from rumors? How was this possible?

"None of that matters now. She's the one. I know it." Micca's voice dripped with conviction.

"We will see."

After a couple of minutes, Larana meekly emerged from the office. She'd gotten the robe on, but the coiled lengths lay somewhat akimbo. Her face turned red as the two men eagerly looked her way.

"Here, let me help." Micca approached her with a winning smile and tucked the lengths away properly. Larana's blush deepened as he helped her, but by the time he was done, she gave him a return smile.

"Thank you."

Micca's smile grew even brighter.

"Sit here, please, Larana. Facing to the side, if you would." Rux stood and brought over the chair he'd been sitting in. He set it close to a lamp so there would be plenty of light.

She did as she was bidden. Small gasps were heard from the two men as her marks were revealed.

"I mean no disrespect, but I will need to touch them. Is it all right?" Rux asked quietly.

"Yes." Larana's answer was little more than a squeak.

The ambassador's hand shook as he reached to touch the birthmarks. He traced their outline with his finger, his eyes growing wider by the moment.

"El is a kind god." Rux stood back his face filled with awe. "There is no doubt, you are El's Vassal. Thank you for coming back to us."

He took Larana's hands; and then, as one, both he and Micca dropped to one knee before her.

Torren watched all this from his place by the wall, noting the flush of embarrassment coloring Larana's face at their deferential behavior.

"Micca, I know it's late, but the council must be told of this at once. With things as they are they will need to wait until tomorrow to see her, but they will at least be able to begin preparations for her return."

Micca gave a quick nod. "Leave it to me."

He bowed in Larana's direction, his bright smile back on his face, before rushing from the room.

"Are you hungry? Thirsty? I can have the cook prepare you anything you wish."

Larana said nothing, looking completely overwhelmed. "I–I'm fine."

"You'll have the use of my room for this evening. It is in

the Lander style, but not uncomfortable." Rux's wings fluttered back and forth.

Torren pushed away from the wall. "I'll leave her to your care, then."

Two pairs of eyes stared at him in startled surprise, the ambassador's face then darkening with suspicion. Torren looked at neither one as he headed in the direction Micca had gone, not caring what any of them made of his abrupt departure. He'd finished doing what he'd set out to do. It was time to get away from this place and get on with his life.

"You can't!"

He half-turned at the panicked protest. Larana rushed toward him, almost tripping over the too-long robe. "Please stay. Only one more day. Please." Tears filled her eyes as she pleaded with him in a whisper, "You're the only person I know here."

Her obvious need bothered him, though he tried not to show it. The sooner he cut away from her the better it would be for both of them.

Rux stepped forward, standing protectively behind Larana. "It...It is somewhat unusual, but if the Vassal desires you to be here, you're welcome to stay."

Torren regarded the tense lines on the man's face and realized how hard this had been for him to say. It spoke volumes of the need the Chosen felt for the Vassal that Rux would even consider allowing a grub to stay on what was, for all intents and purposes, land belonging to the Chosen of El.

Letting out a slow sigh, he gave in.

"As you wish."

Larana almost collapsed from relief. "Thank you."

"Vassal, let me show you the room. I'll have your clothes and other items brought up to you there." Rux opened the door to his office. "This way."

The office was large and filled with shelves full of books and scrolls. A number of large maps were affixed to the walls, pins and lines covering a lot of them. Two chairs faced a richly polished desk, a painting of the Chosen's ascension covering the back wall. Rux crossed the room and opened a small door on the left. It revealed a short hallway and a flight of stairs.

The ambassador led the way, Larana following, with Torren bringing up the rear. Her foot caught on her robe, and she flailed backwards. Torren caught her before she could fall.

"I'm so sorry." Larana scrambled to hitch up the robe so it would trip her no more. He held on to her, her embarrassment, gratitude and trepidation radiating through their connection. He couldn't help but notice Rux's nervousness at the extended contact.

The stairs opened up into a wide hallway on the second floor. Columns and realistic paintings of blue skies lined the way. At the far end was a carved door. It was here Rux led them.

He opened the door and stepped to the side. "I hope you'll find it satisfactory. I will send someone back to let you know where your friend will be staying once a place has been made for him."

He didn't look in Torren's direction, his back stiff, wings bunched together.

Larana stopped in the doorway and stared at the floor before turning around and looking the ambassador bravely in the face. "I...would much prefer it if he was able to stay here with me."

Rux blanched. Torren felt the sudden urge to chuckle but kept his amusement in check.

"He won't hurt me. He's protected me and has gone out of his way to get me here. Without him I would never have found out I was the Vassal." Larana spoke in a rush before Rux could try to deny her. "I would...I would feel a lot more comfortable knowing he was here. Please."

The ambassador studied her a moment then glanced at Torren, his eyes veiled. He finally nodded, it being only too obvious he didn't trust himself to speak.

"Thank you!" Larana reached out and gave him a hug. His eyes widened as their skin touched. Larana backed away after a moment, her eyes searching his. "Goodnight."

Rux's face noticeably softened. "Goodnight. I'll send someone to wake you in the morning and bring you breakfast. There'll be some people here then who'll be very eager to meet you." He gave her a smile. "It is truly a miracle you're with us again."

He made a slight bow to both of them, turned and went back the way they'd come.

Torren followed Larana into the ambassador's bedroom. A small bed was set in the corner, not taking up much room. Tall windows took up the back wall, giving a darkened view of the garden below. A nook in the spacious room held a less ostentatious desk than the one downstairs that was littered with papers. A door close to it led to a dressing room while another open doorway revealed a large bathroom with a tall-standing tub built to accommodate wings. A round table with chairs sat close to the door.

Larana walked from one end of the room to the other, staring at all there was to see. He grabbed a well-padded chair and dragged it closer to one of the lit braziers, then sat down.

"Torren, what's this?" She was examining the small bed. At the top it was wide but then narrowed down through the middle before spreading out again like an hourglass. The upper third was set at a slight incline.

"It's a Flyer bed."

She glanced over at him. "A Flyer bed?"

"Yes. They don't enjoy sleeping in the same types of beds we do—the wings force them to sleep on their stomachs, and most prefer to sleep on their backs. Only on a bed like this can they do so."

"Oh." She tried lying down on it and almost fell off trying to balance herself on the thin backing. "This is hard." She sat up. "Will I have to sleep on one, too?"

"No, not if you don't want to," he told her. "As the Vassal, you can have any kind of bed you want."

Larana blinked several times. "I see."

He stood up at a soft knock on the door. Opening it, he found one of the bronze-armored guards with two others behind him.

"The ambassador asked us to bring your possessions. We've brought a cot as well."

The men's stance was tense and stiff, their eyes hard.

Torren said nothing but moved out of the way so they could come in. The tension was thick as they brought Larana's clothes, his pack, a cot and several blankets.

The girl watched; and as they turned to leave, she gave all three a shy smile. "Goodnight."

They bowed curtly to her but said nothing. The last sent a hateful glare in Torren's direction. He shut the door behind them.

"Why are they angry?" Larana's voice sounded suddenly small.

He shrugged, grabbing his pack and checking through it. The contents were all there except for the dagger he normally kept in a side pocket. None of his other weapons had been returned, either.

"Flyers don't like Landers."

Her small brows bunched together. "Why?"

He shrugged again. "They have wings, Landers don't. They were chosen, others weren't. It's forever been this way."

"I don't have wings. Will they look down on me?"

He glanced at her, sitting back down on the chair and covering up with a blanket. "You're the Vassal, not a Lander. It's not the same."

"Why?" she asked, her face perplexed.

"It just isn't."

Larana walked slowly to the cot and sat down, gaze locked on her hands, clenched in her lap.

"You'll be safe with them. You have nothing to worry about."

"Yes, I know, but..."

Torren frowned. "But?"

She sighed heavily. "All I've ever been is a farmer's niece. I know almost nothing about the world, El or the Chosen, and even less about being his Vassal." Her large eyes turned toward him, revealing her doubts. "They... They seem so happy to have me here, so relieved, but I have no idea what I'm supposed to do. I'm going to disappoint them."

At times, it had been said the Vassal communicated directly with El—that he or she innately knew about their people, their god. Yet this girl held no such knowledge, had no such communications. He'd seen the shock in the ambassador's eyes as he, too, realized this. To Torren, it was only more proof of certain beliefs he'd come to accept.

But he told her none of this. "You'll learn. And no one is expecting anything right now—they're just glad you're back. What matters is that you'll be safe."

She didn't look too reassured. Torren's frown deepened.

"It's been a long day, and you're tired. I'm sure things will look better in the morning."

Larana nodded, not looking at him, and reached for her clothes. She went into the bathroom to change. When she came back, he pretended to be asleep in the chair so there would be no more questions. He wasn't the person to still her fears; he had too many problems of his own.

An hour later when he was still unable to fall asleep, he opened his eyes and glanced across the room toward the cot. Larana was curled on top of it, appearing small and alone. He looked away.

But he'd actually done it. He'd found the Vassal and returned her to her people, though he'd despaired of ever being able to do such a thing. It was over. He'd be able to get on with his life and make of it what he could.

Now all he needed to do was figure out what that would be...

# CHAPTER 13

A SOFT KNOCK AT THE DOOR JERKED TORREN AWAKE. Pushing to his feet, he rubbed at his stiff neck and hobbled toward the door. He glanced back and noticed Larana had yet to awaken.

Opening the door only wide enough to peer into the hallway, he found a grinning Micca waiting there. The young guardsman held a food-laden tray and was no longer wearing armor.

"I've brought breakfast."

Torren nodded and opened the door enough to let him in. Micca set the tray on the table by the door, his bright eyes moving to the occupied cot.

"How is she?"

He was slightly amazed to find the ambassador's nephew willingly speaking to him. He could only assume the joy at having the Vassal returned had momentarily set aside his normal abhorrence of grubs. It wouldn't last.

"Still sleeping."

Micca nodded, still staring toward the bed. "I haven't closed my eyes yet. I've been afraid of finding out it's only a dream." He grinned, finally turning to Torren. "My people owe you a great debt."

So, that was it. He was amazed the Flyer didn't consider the imagined debt galling, having the Vassal found by a grub instead of one of their own.

"I'll be bringing some new garments for her. The Council of Elders has sent some representatives to do the final check, and they are all very eager to see her." Micca glanced once more in Larana's direction.

"I'll wake her," Torren volunteered. It wouldn't be long before their paths would deviate once and for all.

"Wait." The young Flyer's gaze remained on the sleeping girl a moment longer before he reluctantly turned to Torren again. "What can you tell me about her? What are her likes, dislikes?"

Torren found the second question a little odd. "I haven't really known her all that long. Why do you ask?"

His cheeks colored slightly. "No reason. Just curious." Micca turned toward the door. "I'll be back before long."

Torren followed and closed the door after him. Yes, it'd soon be over. He just wasn't finding as much pleasure in the thought as he'd expected.

Figuring he'd best get on with it, he strode over to Larana's side. He studied her for a moment, the loose strands of hair from her braid framing her peaceful face.

"Larana?" He got no reaction. "Larana, it's time to wake."

The girl moaned softly and half-turned from him. He reached to shake her lightly. The moment he touched her, her eyes snapped open. Panic filled them for a moment, but the second they met his they cleared.

"Torren."

"Breakfast is here."

As soon as she spied the tray full of cut fruit, rolls, butters and jams her face lit up; it was definitely better fare than she'd had since the two of them left the caravan. She got out of bed, eagerly reached for a roll and smothered it with a heavy layer of honey butter, then popped a slice of apple in her mouth.

"It's so sweet!" She grinned from ear to ear then bit into her roll with relish.

She ate ravenously, but Torren only took one roll with a bit of jam. Strangely, he wasn't hungry, though they had eaten little the day before.

Larana had just finished, sighing with contentment, when there was a knock at the door. Torren answered it and moved aside to allow Micca entry. This time he'd not come alone. Behind him, staring at the floor demurely, was an old woman carrying a small chest. Her wings drooped slightly, a yellowish tinge at the roots of her feathers.

At the sight of them, Larana went still. Micca dropped to one knee, followed closely by the old woman.

"Good morning. I trust you slept well?" The joyous grin he'd greeted Torren with was back on his face.

Larana glanced quickly at Torren, as if at a loss, then nodded. Micca and the old woman stood up.

"This is Luta," he said. "She'll be helping you dress, if it's acceptable?"

Her next glance in Torren's direction lasted longer. "I…"

She bit her lip then nodded again.

Without a word, Luta headed toward the bathing room, still carrying the small chest. Larana hesitated a moment before following her.

Torren leaned against the wall, prepared to wait. Micca said nothing, meandering over to the table and filching some of the leftover fruit.

"You said you've not known her long…" he said after several minutes went by.

Torren wondered what this was about. "No, not long."

"How did you…?"

"Find her?" He looked away. "It was an accident. If not for the men pursuing her, I might have never known."

Micca's eyes grew big. "Pursuing her?"

Whatever else might have been said was lost as soft giggling trickled into the room from the bathing area, distracting them both. Torren straightened, as if the sound had snapped him out of a half-dream. What was he doing? He'd finished what he'd set out to do. This would be the perfect time to leave. Larana was safe; she was being kept busy. He could avoid a long farewell and tears. He was sure none of the Chosen would object to his leaving. All he needed to do was pick up his pack and go.

But now that the thought was there, he hesitated. He'd left her once that way. On that occasion, he'd been forced to come back. She'd not taken his leaving well. This was supposed to be a time of joy and discovery for her—he'd ruin it by just leaving. He'd go, of this there was no question. This time, though, he wouldn't just run away.

"I'll be back shortly."

Torren had forgotten for a moment he wasn't alone. The ambassador's nephew picked up the emptied tray. Torren's thick brows drew together as he thought he detected a strange expression partially concealed on the other's face. He wasn't sure what to make of it.

When Micca returned the look was gone. Not long after, the door to the bathing area opened.

Luta came out looking as serene and serious as before, until she glanced at Micca. Then her eyes seemed to get a life of their own and a small smile dabbed at her face. Torren totally dismissed her from his mind as Larana came into view.

Startled, he could do nothing but stare. Gone was the farmer's niece with the banged-up knees and disheveled hair. She'd been replaced by something astonishing.

Unlike the robe she'd worn the night before, Larana's new garment was a startling white. The folds and loops were wrapped correctly, and it fit her to perfection. Her hair had been loosened from the eternal braid and cascaded in waves behind her, held back from her face by a golden headband. Her face looked scrubbed and fresh, a touch of color augmenting her cheeks and lips. Golden armlets adorned her upper arms, and there were rings on her fingers.

Before him stood a person worthy of the title of Vassal.

Micca went down on one knee at her entrance, sending a glowing grin in Luta's direction. Then his eyes were only for the girl before him.

"You look wonderful."

Larana held her hands self-consciously before her and blushed. It made her look lovelier still and hinted at what beauty might be hers in the future. Torren finally managed to look away.

"If you're ready, Aen, we should go present you to the councilors." Micca rose to his feet, all smiles.

Larana looked up for the first time. "Aen?"

Micca nodded. "Forgive me, but it's who you are, without a doubt." He saw she was still confused. "All Vassals are given the name Aen. It was the name of the very first of their kind. It was the name you were given when you were born. Unless you would prefer your other name?"

"Oh...either is fine," she said quickly. "I'll try not to forget it."

Micca chuckled, looking embarrassed and reassured at the same time. "I'll remind you, if you wish."

Larana nodded.

"This way, then, please." He opened the door and waited.

For the first time, Larana looked at Torren. She blushed even darker than before and stepped forward. He didn't meet her gaze, instead moving to retrieve his pack. Luta fell in step behind the girl, and he brought up the rear.

Micca led them to the stairs they'd used the previous night. He offered his hand to Larana to help her down.

Larana hesitantly took it and descended slowly, as if afraid she would stumble at any moment. Torren couldn't help but think what an astonishing entrance that would make.

When they entered Rux's office, several people were waiting for them. All eyes turned in their direction as a hush settled over the room. Torren stayed in the stairwell, half-hidden in shadow and out of the way.

Rux hurried forward and knelt before Larana then stood, took her hand and escorted her to the others. There were three Flyers, all wearing white robes with a purple stripe on the hem of the sleeve and golden wings on the shoulder marking them as members of the Chosen's council.

"Gentlemen, as promised," Rux began, "the Vassal of El has been returned to us."

The three Flyers stepped eagerly forward as Rux turned Larana so her back was to them. The oldest of the three, his face pruned with wrinkles, shouted with wonder, his wings making small, flapping jerks behind him.

"It's true!" He immediately dropped to one knee. "El has not forsaken us."

"I don't believe it, though it is here before my eyes." The second councilor, his back stooped, also went down on one knee.

The last man, younger than the other two, stood with

126

his wings straight back, an unreadable expression on his face.

"What more proof do you need, Tel Valerian?" Rux asked as the last did not move to kneel. He guided Larana to face the men again.

Valerian remained still a moment longer, and then a subdued smile crossed his square face as he, too, dropped to one knee. "Welcome home, Aen."

Torren watched as Larana tried her best to look brave. "Th–Thank you."

Now they would whisk her away. Her days of toil and trouble were over. Homespun clothes would be replaced forever with fine robes. If she continued tripping over everything in sight, there would always be someone there to catch her before she hit the floor. Her every need would be attended to. The waif he'd traveled with would soon be nothing more than a memory.

All three councilors stood up.

"So, tell us, Aen," Valerian prompted. "How is it you've been returned to us?"

The eyes in the chiseled, handsome face stared into hers as he towered over her.

"Sir," she said meekly, "it was my friend's doing. He knew what the markings on my back meant and knew where I needed to go."

"Friend?" Valerian stared questioningly at Rux. Then he registered Torren's presence in the background for the first time. "A grub?"

"What? Where?" The oldest of the councilors stared about in confusion until he saw where Valerian was looking. The stooped councilor took a shaky step back. "Dom Rux, what is the meaning of this?"

Torren surveyed them, not at all surprised by their reaction.

"Please, gentlemen, calm down," Rux pleaded. "He's the one who brought the Vassal to us. Without him she wouldn't be here. And it was at Aen's request he remained until the council's acceptance of her identity." The ambassador's wings quivered with agitation. "Surely, this makes his presence acceptable."

The two older councilors reluctantly conceded the point,

though they wouldn't make direct eye contact with Torren. He wondered if they'd ever even seen a Lander up close before.

"And what, exactly, would make a Lander give the Chosen such uncharacteristic charity?" Valerian's question was aimed directly at him. "Could it be you're expecting monetary compensation, knowing the high worth we place on the Vassal?"

His sarcasm was heavy.

"Valerian!" Rux stared from the councilor to his unusual guest and back again.

Torren clenched his jaw, his gaze not leaving the councilor's face.

"Speak up," Valerian taunted. "Have you nothing to say, Lander? Is the truth too obvious to be said?"

"Stop it!"

All turned in amazement to stare at Larana as she interposed herself between the two men. "You...You shouldn't do this. He's not one of them. He's one of you!"

Rux spun to look at him, but Torren only stared at the floor, his heart pounding. Was she guessing, or did she actually know? Though he'd not said anything about it to her, he realized the evidence had been there for her to draw her own conclusions.

He hadn't counted on this.

"My dear," Valerian said, stepping forward to sweep Larana into his arm and pull her away from Torren as if she were a silly child. "It's true he has our coloring, but he has no wings. Only the Vassal grows no wings amongst the Chosen." His hard gaze returned to Torren. "At most, he might be a half-breed, begotten by foul means. And as such, and having been raised amongst them, he is one of them."

Larana seemed to almost wilt beside the councilor as he held her, half-covered by a wing, as if he were claiming her for his own. With a shiver, she forcibly pulled away, shaking her head.

"You don't understand," she insisted, looking at the others. "He is one of you." She turned to him. "Torren, tell them. Make them understand."

He shook his head slowly. Let them think whatever

they wanted. It was time for him to leave.

"Torren?" Rux was studying him with sudden intensity. "As in the son of Lar?"

His skin went cold. Yes, it was high time he took his leave.

"I should go." He stepped out of the stairwell, his pack in tow, not looking at any of them. As he emerged into the light, Rux moved in front of him, his scrutiny sharp.

"El be praised, it is you." He shook his head in wonder. "I can't believe I didn't see it sooner."

Before Torren could stop him, the ambassador grasped him in a fierce embrace.

"It's a miracle."

"Rux, explain this! Why are you further soiling yourself with this grub?" The stooped councilor tapped his cane against the floor, making a ringing sound on the marble tiles.

Feeling incredibly tense and awkward, Torren was only moderately relieved when Rux finally let go of him. This man...knew him?

"Look at his face, Tel Icos," Rux demanded. "Don't you see? This is Torren, Tel Lar's son. He accompanied his father when Lar went in search of the Vassal after her disappearance."

Torren forcibly refrained from taking a step back as the three councilors stared at one another in confusion then seemed to move forward as one to take a closer look at him.

"By He who guides us, there *is* a resemblance." The oldest of the three men peered at him, wings drooping.

"But, Mides—"

"Hold, fellow councilors," Valerian interjected, holding up his hands. "Let the fellow speak for himself. Let him explain how it is things have come to be as they are."

Torren threw Valerian a look from the corner of his eye, not sure from the Flyer's expression whether he truly wanted an explanation or was just trying to give him enough rope with which to hang himself. When he said nothing, Valerian's face acquired a knowing smile.

"Torren, please tell them." Larana stared at him from where she stood slightly off to the side, her eyes troubled.

"They need to know."

He glanced over at her, at her earnest face. Explain? He didn't want to explain. What was there to explain? Wasn't the fact the Vassal had been returned all they should care about? Why not leave other things alone?

He almost jumped as a hand fell on his shoulder.

"Please, do us this favor," Rux said with deep emotion. "Your father was my friend, some of those who went with him my relatives. We need to know what became of them."

Torren's hands clenched and unclenched at his sides, his eyes were locked to the floor. After several moments, he gave a long sigh and nodded. He could feel Larana standing by him, worried; but he didn't look at her, instead crossing to stand before a brazier on the right.

"I doubt I'll be able to tell you much."

"Whatever you can will be fine." Rux's voice was kind.

He hesitated a moment longer, not having thought he'd ever have to do this. He forced himself to start.

"When the Vassal disappeared, my father, as many of the other councilors with experience below did, gathered a group of men and set off in hopes of picking up the trail of whoever took her. We'd been traveling for many days, questioning every Lander we met. No one seemed to know anything, but we were determined..."

*There'd been fifteen of them—his father, some aides, cousins, friends and Torren. His father had served as an ambassador in one of the other countries for a time and had dealt with Landers, unlike so many of the Chosen. It was logical he would be one of those who went looking for the missing child.*

*He had been so proud when his father allowed him to come along. So pleased he could do his part to help El and also get a look at those who lived below—the barbarian Landers. He'd even been given the duty of carrying El's standard, and he'd done it with pride. He'd been such a fool.*

"Eventually, we came across a farmer who told us of a group of strangers he'd seen in the early morning, six days or more before. We went where he instructed us, into a small ravine with rock overhangs, straight into an ambush. It was close to dusk, and before we realized what was happening, nets came down on us from above, trapping everyone.

"Though we tried to fight and escape, we couldn't get free. The attackers came and cut us down."

Torren closed his eyes, the memory of his companions' agonized screams echoing inside his head. His father turning to him, trying to push him as far from the attackers as he could, and getting a sword in his back for the trouble. And they'd died for what? For *what*?

A soft touch on his hand made him open his eyes. Larana stood beside him, her eyes reflecting his anguish.

"They massacred everyone."

Silence permeated the room. Then Valerian spoke

"Yes. Well, I don't mean to sound insensitive, but how is it you managed to survive?"

Torren felt his whole body stiffen. He didn't like what the arrogant councilor's tone implied.

"It was only due to luck, fate—whatever you want to believe in. I'd been pinned beneath my father's body. When they thought we were all dead and started removing the bodies to a pit they'd dug nearby, they discovered me. They decided they wanted to have some fun."

His back twinged, his face clouding with the memory of agony.

"After they were done, they threw me on top of the others and half-buried us. They left me for dead, but I was able to make my way out of the loose dirt." His voice grew quiet. "A farmer and his wife found me on the road near dawn. They carried me away with them. They treated my wounds, and it's only because of them I survived."

His throat was dry, the hatred and the gratitude he felt over their kind act still clashing inside him after all this time.

"So, that was how you..." Micca's face was pale, his eyes wide. His wings were folded around his body as if to protect him from his own thoughts.

Torren said nothing.

"Can you tell us anything about who attacked you?" Rux asked.

He shook his head. "Not really. They were Landers, but they wore no identifying marks or insignia. They knew what they were doing and were thorough about it. The attack wasn't random; they weren't bandits. They'd been waiting for us and knew how to incapacitate us to take away our air advantage."

"In other words, your father got too close," Valerian mused.

Torren looked up surprised.

"None of the other groups sent out to search for the Vassal were harmed but yours," explained Tel Icos.

He frowned, not having ever made that connection. Could they truly have been so close? For the first time, he realized the area where Larana spent her youth was only a day or so from where the attack occurred. He'd been so close all this time and not known it? His nails bit into his hand.

"So, we have Aen safely back, but still have no idea as to how she was taken or who took her." Icos dropped into a chair. "It might happen again."

"It won't!"

Everyone turned to look at Micca. The young Flyer moved to stand by Larana.

"We've searched too long for her. Too many lives have been lost. Too many have suffered. I for one will not allow anything to happen to the Vassal again. I swear it."

One of his wings unfolded partway to hover protec-tively behind her.

"You're right," said Valerian. "Aen will not leave us again." The councilor's face was hard.

Torren stepped away from them. "My time here is fin-ished, so if you'll excuse me."

He started toward the double doors leading out into the hallway.

"But..." Rux took a step after him as if to intervene but didn't.

Torren had almost reached the doors when a familiar voice did bring him to a stop.

"Wait!" Larana deftly dodged past Valerian and the others and ran to stand behind him. "Torren."

Though it should have been simple, he couldn't bring himself to turn around and look at her. He knew what she wanted, but he couldn't—wouldn't. His old life was gone.

She couldn't possibly expect him to take it up again, not as a cripple.

"Yes?"

She said nothing, instead reaching out and taking his hand. Apprehension and loss flooded through him, and though a little of it was his, most of it belonged to her. But why should she feel this way? They hadn't known each other long. And she had nothing to be apprehensive about; she had a new life, comfort and safety in her future.

He half-turned at last. She looked over her shoulder at the others as if to assure herself none would come near. When she turned back to him, he realized she was afraid. And it wasn't fear of losing him but something deeper.

"Larana?"

"I just...I just wanted to thank you again," she told him quietly. She put something in his hand. "And to give you this. It's the only thing I have that's my own."

He was tempted to look down to see what it was she'd given him but didn't. Couldn't. He was trapped by her eyes as tears formed in them but didn't fall. He said the first thing he could think of, though he didn't understand all the connotations of it himself.

"It's been an honor to know you."

Larana's eyes brightened, the fear momentarily gone. "Please be happy."

He gave her a half-smile and a nod. "You, too."

The next thing he knew, he'd turned around, opened the door and left her.

As he stepped out of the building into the morning sunlight, Torren felt certain he was leaving his old life behind finally and forever.

# CHAPTER 14

S EVEN DAYS. IT HAD BEEN SEVEN DAYS SINCE HE'D LEFT her, and they felt like the longest days of his life.

Torren drank the rest of his ale, not tasting it. His eyes locked again on the blue hair clip Larana had given him as he slammed the cup down on the table with a dissatis-fied sigh. He barely noticed the curious looks it earned him and even less the worried ones coming from across the bar.

What was he still doing here? What was he waiting for?

After he left the ambassador's home, he had ended up at Sal's door rather than on his way out of the city as he'd planned. Sal had been only too happy to see him and had wined and dined him as he told his merry tale about how the expected attackers had come, and how, just as they'd started banging on his door, the watch had shown up. The guards had driven the assailants off in a rout, smashing many heads while they were at it. Sal laughed as he told Torren how his customers had gotten a little more for their money than usual.

He then asked about Larana, but after Torren's half-hearted, incoherent reply, he'd not asked again. Torren went to bed that first night, his head ringing from mead and wine, planning to leave the next day to pursue his original plan of going north.

He never did, for that same night he dreamed his old dream of horror and pain, reliving the day his father and the others were killed, the occasion during which he had lost his life as a Chosen—the day on which everything he was and believed in was destroyed.

But this time—this time the nightmare was different. This time he was not the one carrying the standard, not the one watching helplessly as others were mercilessly butchered. It was not he who was dragged from the nets, poked and shoved and, finally, filled with agony as his wings were sawed from his body before he was dumped to die like an animal. This time, it was Larana who suffered all this—and more.

Torren awakened feeling cold and weak. In the darkness, all he was able to think about was the apprehension he'd felt from her, the fear he thought he'd seen in her eyes as he turned to leave her. But what did one have to do with the other?

Yet, though he possessed no answer to this question, the dead certainty the two happenings were related wouldn't leave him.

When the sun had risen and light once more intruded into his world, he found the floating city of the Chosen gone. He'd wanted nothing more than for this to be so, but now it brought him no satisfaction, only dread. All he could think about was his transformed dream and the reasons it might have changed in such a strange and horrifying way.

That first day he had sat where he sat now, uselessly pondering the question, worried over he knew not what. He had the new dream again, and this time he saw it was his name on her lips as she screamed in agony.

Again, he'd not been able to sleep after the nightmare, so dawn and Sal had found him back at the corner table. Sal asked him what was wrong, but Torren only stared at him and said nothing until his friend finally went away. How could he even attempt to explain something he himself didn't understand?

He thought of leaving, of running away. Perhaps a change of scenery would make the nightmare disappear. But something inside told him he was where he needed to be. If he hadn't known better, he would have thought he was acting like a lovesick pup mourning the loss of the object of his affection.

In a different way, though, he felt almost driven to stay, as if someone were trying to tell him something through his dreams and all he had to do was figure out what. It was ludicrous; but as infuriated as doing nothing made him, he trusted his instincts. So, he sat and waited, staring at Larana's gift as, night after night, the dream returned and he grew more and more isolated from everyone and everything around him.

# CHAPTER 15

"BY THE GODS, TORREN, HASN'T THIS GONE ON LONG enough?" Sal whispered harshly as he set a food-filled plate before him. "Talk to me, you stubborn fool."

Torren only stared at the hair clip.

"You hardly eat, I doubt you sleep. You only sit here brooding, scaring off my customers." Sal stared at him anxiously, rubbing at his scar, his words falling on deaf ears. "Torren!"

His friend grabbed his arm and shook him, looking for some kind of reaction.

Torren finally gave him an angry glare. "Would you prefer it if I left?"

"Dammit, no!" Sal waved his arms in frustration. "It's just that I've never seen you act this way. Not even when I first met you, and you were such a disagreeable and angry young man. I only want to help."

He shook his head. "There's nothing to help me with."

"But, Torren, surely...Does this have anything to do with the lass?"

He tensed, his hands curling into fists on the table. He suddenly turned to look at Sal, his eyes intent. "Do you believe in the gods—I mean, truly believe in them?"

Sal stared back, caught off-guard by the question, then gave his friend a half-shrug, not sure where he was going with this but happy to have him talking.

"I don't honestly know if I do or not. They've always been a part of my life, one way or another, and I've never been one to take unnecessary chances. Why?"

Torren only shook his head, unable to explain. He could

see Sal was about to press him when they were interrupted.

"By all that's sacred, I'd hoped you'd be here."

Torren turned, startled by the familiar voice, and spotted Micca hurrying toward him. He frowned deeply, not at the Flyer's unexpected yet somehow unsurprising arrival but at how he was dressed.

Micca wasn't wearing either the long or short robes that were the Flyers' customary garb. Instead, he was wearing pants and a loose shirt similar to Torren's, as well as a large cloak and a backpack. While his wearing Lander clothing was surprising, what was more so was that he was using the backpack and cloak to disguise the fact he had wings.

"Torren, I must speak with you."

His heart pounded in his breast, his skin felt prickly, a dead certainty stealing over him this was what he'd been waiting for.

"You know this man, Torren?" Sal asked, eyeing the newcomer warily.

"It's urgent," Micca insisted, ignoring Sal and leaning over the table. The Flyer's face was haggard, as if he'd traveled many days without rest or sleep.

"His name is Micca," Torren said in answer to Sal's question as he surveyed the inn. It being mid-afternoon, the common room was empty except for them and a couple of old-timers at a far table. His clear gaze came to rest once more on the Flyer. "How did you know I was here?"

Micca shot a suspicious look at Sal then dropped down onto a bench close to Torren. "Larana. She told me of your time together, and how you'd brought her to this inn." The Flyer rubbed his face, exhaustion almost pouring off him. "I–I took a chance, hoping you might be here."

Without being asked, Sal dragged over a chair and sat down as well. Torren half-raised a brow at the older man's protective attitude.

"Then she probably told you about him as well." He pointed at his friend. "This is Sal."

Micca glanced over at the scarred man with a little more warmth in his eyes. "Yes...Yes, she did." He gave him a sitting bow. "Thank you for the help you gave her in the past."

"I'd do anything for Torren here, or the young lass." Sal sat back, his expression clearing, his eyes shining with growing curiosity.

Torren's gaze locked with the Flyer's. "Micca, why are you here?"

His insides were churning. Dread filled him—whatever news Micca brought with him wouldn't be to his liking.

The Flyer hesitated, sending a worried glance in Sal's direction. "I mean no offense, but..."

Torren waved his hesitation aside.

"You can speak freely. He can be trusted." Besides, from the look of him, nothing short of a major cataclysm would move Sal from his seat, and he held doubts if even that would do it.

Micca nodded, though still reluctant, and focused on the worn tabletop. He seemed to study it unseeing until he ran across Larana's hair clip. Tentatively, he reached out as if to touch it but stopped.

"Torren, we need your help." His voice was quiet. "I need for you to come back to the capital with me."

Torren went cold. "I don't belong there. I would have thought this perfectly obvious." He amazed himself with the tightness in his voice. He could feel Sal staring at him, perplexed by his reaction.

"Y–Yes, I understand, but..." Micca's light-blue eyes sought his. "...but something's happened, and we need you. We need someone from the outside who knows our ways."

Apprehension coiled about Torren's stomach and rose up into his chest. "Why?"

Micca looked away, guilt and more filling his features. "I failed her. Despite my lofty promise. I failed her."

Torren's apprehension solidified into fear.

"What's happened?" he asked tightly. Of its own accord, his hand moved across the table and took hold of Larana's hair clip.

Micca faced him, tears shining in his eyes. "She's asleep and won't wake up. We've tried everything, but she won't wake up."

"That's ludicrous, unless she's dead." Sal shook his head, his eyes daring Micca to contradict him.

"I know how it sounds, but nevertheless, it's true. She's not dead, but she won't awaken," Micca insisted.

"And what do you think I can do?" Torren stared at his clenched hand, the clip eating into his flesh. A sleeping death. Why?

Again Micca hesitated. "For...For years, there have been rumors of things no one was willing to admit, but now that disaster has struck again, they've been given more validity." He rubbed his eyes. "It's been a suspicion of some the last Vassal didn't die a natural death. It's been suggested he was poisoned."

Sal's eyes grew wide even as Torren felt his chest grow tight.

*What?*" The death of the previous Vassal had occurred a year before Larana's kidnapping. Though he'd only been a young boy then, he didn't recall hearing that the death had been anything but a peaceful passing due to old age.

Micca's tired eyes held his. "Yes, the unthinkable. But with his death, the kidnapping of the new Aen as a babe, and now with her falling into an endless sleep, the un-thinkable seems very probable."

"But who would do such things?" Torren demanded. "Landers don't have access to the capital, let alone the Vassal."

Micca's face twisted as if in pain. "That's the problem, don't you see? If someone is responsible for these things, if someone purposely poisoned the Vassals, then another unthinkable fact must be true—it was done by one of our own."

Torren sat back, the truth of it staggering him. A Chosen had poisoned the Vassal? It wasn't possible!

"Hold on, here." Sal shifted forward, looking from one to the other. "Are you saying you and Torren...and the lass...are not Landers?"

Micca nodded once, after glancing around the room to make sure no one was watching. Quietly, he shifted his cloak just enough to allow Sal a peek at one of his wings.

Sal's eyes grew even wider. "But—"

Torren cut him off. "Sal, it's a long story, and I promise

to fill you in, but not now." Staring until his friend nodded agreement, he then turned his full attention back to Micca. "I still don't see how this has anything to do with me."

The Flyer's look implored. "She needs you. *We* need you. It's imperative we find out who's behind all this and why. But no one belonging to the council will bring it up—it would destroy our people, even if they were willing to believe such a thing. There are those already spreading the word this is El's retribution for letting the Landers take her in the first place.

"Only someone from the outside, yet someone who is one of us and knows Lander ways, has any chance of figuring out what happened." He hesitated a moment and then plunged on. "And though what took you from us was horrible, it has also put you in the unique position to do what the rest of us cannot. And maybe, if El is willing, you might find out enough to let us save her."

Torren laughed humorlessly in the Flyer's face, the old bitterness rising into the back of his throat. "You don't know what you're asking me."

"You're right, I don't," Micca admitted. "But it still doesn't change the fact your people need you, that *she* needs you, that there's no one else who can do this but you."

His dreams of the last few days. This was what they'd been about. Torren was sure of it. But why? And what in the world could he do?

Again he remembered the fear he'd seen in Larana's eyes, and the anxiety he'd felt from her touch. Did she suspect something back then? Had she felt something from the men gathered around her? And he'd left her there. Left her to be poisoned and sent into an endless sleep after he'd assured her she'd be safe. The signs had been there for him to see, but he hadn't. She'd known, and she'd let him go anyway.

And he'd gone.

He shook his head and opened his hand to look at the clip. The edges had gouged his palm, leaving deep, angry welts. He'd told her she'd be safe.

"All right. I'll go."

Giving the impression he'd been half-expecting a fight, Micca slumped forward with relief.

"Thank you." After a moment, he lurched to his feet, swaying slightly. "We'll need to talk to my uncle first. Then we can start for the capital."

Sal stood up as well, grabbing the Flyer's arm to steady him. "You're in no shape to go anywhere. You're about to drop where you stand."

Micca shook his head. "No, I'm all right. The sooner we get back, the faster all this can be resolved."

"You need food and rest," Torren stated as he got up as well, his tone brooking no argument. "And there are things I'll need to take care of. Eat and sleep for a couple of hours. By then, I should be ready to go."

Micca stared at the two men, looking as if he would argue, but then sighed and gave in, seeing neither would be moved by anything he might say. "All right."

"You can use my room," Torren told him. "Sal will bring you some food." He herded him toward the stairs.

Micca's eyes were already half-closed by the time they made it to Torren's room. It was as if now that he'd received reassurance Torren would help he could allow himself to admit his exhaustion. As soon as his body touched the bed, Micca was asleep.

Torren studied him, marveling at the ingenuity of his disguise and the fact he was even wearing it. Only pure despair would drive a Chosen to this. It pointed to how bad things were. He wondered what he was getting himself into.

He'd sworn years before he wouldn't go back, after he'd finally healed and learned enough to be able to survive in the world and could have returned on his own. Too much had happened, too many things were unresolved. Yet here he was, willingly doing so for a girl he barely knew. He fingered the clip in his hand then placed it in his pocket.

Sal showed up with Micca's meal as Torren placed his pack by the door. Gathering his things hadn't taken long at all. The two men stepped out into the hallway once Sal set the tray down by the window.

"So," he asked cautiously, "you're going?"

Torren nodded, knowing where this was headed and

not sure how to stop it. He wasn't even sure he should.

"And you're a Flyer?" The volume of Sal's voice rose slightly, his hand rubbing absently at his scar.

"Flyers have wings." Torren set off for the stairs.

Sal sent him a startled look then laughed as he hurried to catch up to him.

"It explains a lot," he said knowingly. "You could have told me."

Torren stopped, aware of how much he owed this man, this Lander who, like his saviors, had forced him to live and learn to fight despite how much he'd resented the interference in the beginning. "It wasn't important."

"Actually, I'd always pegged you for a half-breed," Sal admitted.

He stared at him in surprise.

"Well, it neatly explained some of your attitudes and your continuous gloom." Sal grinned. "Besides, it's not as if I couldn't help but notice you never needed to shave. Everyone knows Flyers grow no facial hair."

Torren felt himself blush. He'd never suspected anyone had noticed. For years, he had gone out of his way to mimic the process just so he wouldn't be accused of such a thing. Strangely, he now realized he'd not followed his usual practice with Larana.

He had never met a half-breed in his travels and didn't think they actually existed, but it was a persistent fantasy. The Chosen didn't mingle with Landers unless they were duly elected to ambassadorial posts. Even then, they only dealt with those in authority to trade goods and arrange for transport contracts—bad experiences in the far past kept most of them from wanting to deal with those below. His father had been of a different mindset, but he was in the minority.

The prejudice against Landers went deep, so the thought of a dalliance with one of them was repugnant to his kind. And though in his time amongst them, he had come to find many of what his people held as truths about Landers were false, those in the floating islands fully believed them. Landers were the first to be blamed when anything major went wrong.

"Did the same thing that happened to you happen to the lass?" Sal asked quietly.

Torren stared at his friend in confusion until the meaning of what he'd asked finally clicked into place. "No. The Vassal is born without wings. It is supposedly El's way of reminding us of our origins."

As he said it, he realized none of them truly saw it that way anymore. Though the facts were that the Chosen had once been Landers, it seemed to be something none of them ever really thought about or wanted to remember.

"You'll need food for your journey. Feel free to take whatever you need."

Torren nodded. "Thanks. And if you're willing, I have something you could do for me as well."

Sal agreed without hesitation. "What do you need me to do?"

He said nothing for a moment, descending the stairs and going into the kitchen, striving to put all his thoughts together. "There's a group of men in black leather armor sporting good horses and money. They were the ones who killed Larana's Lander family, the ones who set up the attack on the inn and made a grab for us on the road."

Sal's eyebrows rose at the last, but he didn't interrupt.

"I want you to try to find out who they are and what they're up to. I'm fairly positive they're somehow connected to the rumors about the buildup at the border—that might be the best place to start. It's even likely one or two of them are still here in town. I'll give you what money I have to grease some palms and loosen tongues. If it's not enough, sell my half of this place for whatever you can get for it."

Sal nodded. "I have people who owe me favors from the old days. One or two of them even work for the guard. They might know more about what's going on than they're telling and might also be interested in this band of black-dressed men of yours." He slapped him on the arm. "You can count on me."

Torren looked away. "I really appreciate all this. I wasn't certain how you'd react once you knew the truth. To some, Flyers are as unwanted as Landers are to them."

Sal laughed. "You are who you are, Torren, and it has

nothing to do with where or to whom you were born. All that traveling around at least got that much through my thick skull. All I care to know is you're my friend and you need my help." Torren found himself in a steel-armed hug. "When you saved my life I didn't care what you were. Why should I start now?"

"All right, old man, enough!" He felt a half-grin tugging at the side of his mouth as he tried to pull away. He was amazed how relieved he was that Sal didn't hold his birth against him—he couldn't say he'd done the same.

Landers had killed his father, and he'd been forced by circumstance into looking the same as them, to live among them. He'd picked his profession so he'd have a chance to kill some of them, to get what vengeance he could. But over the years, though he'd not once been reconciled to what had been done to him, he'd learned not all of them were at fault, and that some could be trusted to the point of being called friends.

Sal let him go, smiling widely, slapping him on the back. "Let's get you those provisions."

Several hours later, Torren headed back up to the room to wake Micca. Sal came behind him, carrying hot meals for the two of them to add to what he'd left in the room before. Micca stared at the strange fare warily for a moment, but then dug in with abandon. Though Torren felt far from hungry, he made himself eat as well.

Halfway through the meal, he got up, bent over his pack and brought out his pouch of money. "Here, Sal. Make it go as far as you can."

"What is it for?" Micca asked between mouthfuls.

"Information," he told him. "I don't want to rely only on what can be learned at the capital."

"How will I get what I find out to you?" Sal asked as he tucked the pouch out of sight.

Micca answered before Torren could.

"Contact our ambassador. His name is Rux. He's my uncle and can be trusted. He'll make sure whatever you learn gets to us. I'll let him know you'll be by."

Sal nodded. "All right."

They finished eating. Getting up, Torren grabbed his backpack and sword belt and put them on. He stared at the two men with him, knowing only disturbing times were responsible for such an unusual gathering—a Lander, a Chosen and one not quite either. But it didn't feel wrong. If anything, it felt almost natural.

"We'd better go. It'll take time to get to the embassy, and I want to make sure we don't pick up any observers on the way."

Sal followed them to the door. "May I see you both soon." He placed a hand on each of their shoulders.

Micca appeared startled by Sal's gesture but said nothing, only nodding in response.

Torren opened the door. Warm sunshine blinded him for a moment. Normally not one to stay indoors, he realized with a start he'd not looked at the sun or sky since he'd left Larana. It was almost as if he'd stopped living until the moment Micca came through this very door.

Shaking the thought away, he stepped out into the street. With the Flyer in tow, he took a convoluted route to the embassy. Though chances were no one was keeping an eye on him, he felt he'd taken too many things for granted before to allow himself to make any mistakes now.

The sun was low to the horizon when they finally reached the gated entrance. Micca rang the bell and banged against the locked doors insistently until one of the guards came to check.

"It's me, Micca. Let us in."

The guard opened the door quickly but then hesitated, surprised by the Flyer's unusual garb and his unexpected companion. "Sir, didn't you...?"

"Yes, but I'm back now. I have to see my uncle immediately." Something in his tone startled the guard enough that he hurriedly stepped aside. Micca rushed through with Torren right behind him.

Without preamble, they entered the building and headed straight for the ambassador's office.

"Uncle, are you here?"

"Micca?" Rux stood up from behind his desk, a shocked look on his face. "What are you doing here?"

His brow rose another notch as he spotted Torren behind him.

"The unthinkable has happened," Micca told him. "The

Vassal has been driven into a sleep from which she cannot wake."

"*What?*" Rux's face drained of color. He reached behind him for his chair and sat down. "How did this happen?"

Micca shook his head, not looking at his uncle directly. "We're not entirely sure. She was doing fine—learning the ceremonies, walking amongst the people. She was never alone. Yet after the evening meal three days ago, she collapsed. We've not been able to wake her since."

Torren's eyes narrowed. "How long after the meal was her collapse?"

"She'd just excused herself."

"Did anyone check the food?"

Micca stared at his feet, a hot flush coming to his cheeks. "We didn't get the chance. When we took Aen to her room and sent someone to summon the healer, all we were thinking about was her. It wasn't until later it occurred to me to check it. By then, the dishes were already cleared and the leavings disposed of."

"Was she dining alone?" Torren asked.

"No. That night, Aen was dining with several members of the council. They were discussing the schedule for her visits to the other islands. The one where her family dwells was to be the first." Micca didn't look at them as he added the last.

Torren tried not to let the information bother him. "Do you remember who was there?"

Micca nodded. "Tel Icos, Tel Mides, Tel Valerian, Tel Mallean, and Tel Symeas."

Rux shook his head slowly from side to side. "It couldn't have been any of them—these men and women are above reproach. To think someone in the council..."

"It's why I felt it necessary to seek Torren. He knows our ways but is an outsider at the same time. He...He should be able to see things more clearly." Micca's desperate expression begged for his uncle to agree.

Rux straightened slightly in his chair. "Yes, it might be helpful."

His nephew sighed with relief. He pulled off his fake backpack and cloak, allowing his wings to stretch out freely behind him. "The capital is on its way back, but I didn't want to wait. Some of the councilors hold the vain hope returning to where she was found will revive Aen again."

Still visibly shaken, Rux nodded. "If only it would work...but it won't. El wouldn't have done this only to undo it so easily; and since we suspect others, they definitely wouldn't find it convenient for her to awaken, whatever their reasons may be for doing this in the first place."

"We're going to leave tonight and catch up to the capital before it gets here," Micca told him. "I wanted to make sure I filled you in before we went."

Rux brushed an unsteady hand over his face. "Yes, yes, every moment counts, but you should wait a while. Going over the wall won't be easy, and the guards are very alert at this time of night. It'll be better if you delay your departure. Besides, this will give me a little time to try to make things as easy as I can for the two of you."

Micca nodded. "Thank you. Also, Torren has already requested the help of someone here in town to look into some men who tried to capture Aen before. We told him he could come to you with the information if any was to be had."

Torren left the office. He dropped his pack in the reception room by a chair, the knowledge he was actually going to go through with this settling over him. He must be mad. He was even surer of it as he felt anticipation sparking deep inside him at the prospect. To be once more in the skies, to see the green gardens, the fountains, the towering columns again. He tried to squash the feeling as thoroughly and violently as he could.

"Torren."

He almost jumped at the touch at his shoulder.

"My uncle said he'll have the guards keep a watch for your friend. He'll make sure any news he has from him is brought to us at once. He said he would also supply him with more funds to finance his efforts and some for our travels as well, in case we need it. He's also writing some letters on our behalf."

"All right." Yes, he was really going to go through with this. He tried to find something with which to distract

148

himself. "Could I ask a question?"

Micca blinked in surprise. "Of course."

"I noticed before you wore silver armor. Does this mean you've seen combat?"

Guard positions for the Chosen were, on the whole, ceremonial positions; they were mostly used as a means of discouraging trouble from Landers or as an honor guard. Micca appeared a little young to have achieved such a rank.

"No, not yet. Not real combat, though I've taken part in many exercises." Micca preened his right wing, not making eye contact. "For the last few years, the council has been encouraging us to learn more about the martial arts so we would be ready if there ever was a need. I've always excelled at sports and have deeply wanted to do my part for El, so as soon as I became old enough, I volunteered for service. Enough others joined later that, with my experience here at the embassy, I was given a higher rank.

"I also had to win a number of competitions. I not only did well enough to gain the rank but also to be selected to be one of Aen's guards when she would eventually return to us."

Torren turned partially away, troubled. They gained rank by competition? He stared at his callused hand, recalling how he'd earned his.

"You've fought, haven't you?" There was a sudden brightness in Micca's eyes he wasn't sure how to feel about.

"Yes, I've done my share." He'd done his share and more. But thinking of this young man doing the same somehow bothered him.

"Perhaps you could teach me. Help me be more effective for Aen." The young Flyer's face sobered. "If I get a chance to serve her again."

"I'm sure you will." At least, he fervently hoped so.

A little over an hour later, the three men met in the ambassador's office again. Rux's color had returned, and there was a fervent gleam in his eyes.

"Torren, things have changed since you were at the capital last, but I agree fully with Micca that your unusual viewpoint could be very helpful." He gathered a number of scrolls and passed them over. "These are for some of the people you will meet. Micca will know who they are. Mostly, they contain signed statements that I recognize you as the son of Lar.

"Our people's opinions of the Landers have worsened since Aen's abduction. If the council hasn't already informed the others of your existence, these will take care of it so there will be no question as to your identity. This will hopefully avoid some potential troubles."

Torren flinched, realizing the necessity of what he was saying but not liking it any more than before. He took the documents and tucked them away without a word.

"Micca, I am tasking you with his protection," Rux said seriously. "I doubt this will be easy, especially if things are as we suspect."

"Yes, Uncle, I will."

Torren frowned slightly, used to only relying on himself. Still, he realized in this situation he might have no choice but to rely on others. Handling people wasn't his forte.

"Before you go, let's sit together for a moment and ask for El's guidance and blessing," Rux suggested.

Torren felt the hackles rise on the back of his neck but didn't argue. If he did, they'd want an explanation; and he doubted that, no matter what he said, it would be something they'd understand. That a Chosen didn't believe in El went beyond blasphemy—it was a virtual impossibility.

Rux led the way and took them to the back of the house, to a modest round room. Intricate, realistic paintings of lush ground and sky covered the walls, as if to give the visitors the impression they were outdoors. Small benches were set in the room in a circular pattern, all facing a pedestal of marble. Atop the pedestal sat two miniature alabaster wings.

Torren stared at the symbol of his people's god and felt nothing. He kept his expression blank as the other two took seats close to the center and bowed their heads, setting their hands thumb to thumb with the fingers held out straight—a representation of the wings before them.

Wings—the embodiment of all El meant to the Chosen. The circular temple signified their unity, and the scenes of nature their freedom to serve. It was the wings that made the Chosen stand out from the rest, that granted them their uniqueness in the world.

Wings.

Torren felt the muscles in his back twinge. Others had lost one or both, this he knew, though he'd never met any. But the loss had occurred through accident, not the willful removal he'd been unable to prevent. Anger rose at the fact he'd lived through the humiliation, and he tried his best to bring it back under control. Why should he be angry? There was no god to blame. They didn't exist.

After several minutes, the other two finished, and they left the chapel. He didn't look back.

"It's late enough now it should be safe for us to go," Micca said. They were back in the ambassador's reception room, and he was collecting the discarded parts of his disguise.

"My hopes go with you." Rux embraced his nephew then clasped Torren's arm. "I'm glad you've come back to us. I just wish it'd been under better circumstances."

"Yes." He didn't agree but saw no point in saying so.

"Be careful, both of you."

They took their leave, picking up what few extra supplies they deemed necessary. Micca led them out the side door, through an archway into the garden and from there to the embassy's back wall. Dressed once more as a Lander, he let Torren boost him up to the top of the wall. Once settled, he held his arm down to help Torren gain a purchase as well. They disappeared down the other side into the darkness.

Keeping to the shadows, Torren led Micca through the streets until they reached the city wall. This was where things would get tricky. To get out, they'd have to go up the wall, cross the open battlements and slip down the other side without being seen.

Most of the city guard held stationary posts, but a few made rounds. Their only advantage was the guards would be looking for trouble from without and not from within.

Keeping watch, he made sure they weren't being observed as Micca quietly removed his pack and cloak. Flexing his wings for a moment, he nodded to him.

"I'm ready."

Picking up Micca's discarded things, he felt his stomach suddenly clench in nervousness. He stepped up to the wall, his back to the Flyer. "All right, let's get it over with."

Micca came up behind and snaked his arms through Torren's, reaching up and over to his shoulders. Torren felt the excited Flyer's breath fall softly against his neck as his wings spread out to either side of them. "Jump on the count of three. One...two...*three!*"

The two men jumped at the same time, but instead of falling back to the ground once the upward momentum was spent, gravity reaching up hungrily for them, they continued moving up. Micca's wings flapped hard. Torren clung to his arms, not having dangled high in the air for more than half his lifetime and finding, at the moment, he wasn't too excited at being there again. The ground continued to drop below them; and after taking a dizzying glimpse of it, he decided to keep his eyes on the wall instead.

Micca's wings made little sound as they propelled them upwards. His grip on Torren was confident and sure.

Their ascent slowed as they neared the lip of the wall. Both peered hard to either side, making sure no one was in sight. Micca carried Torren halfway over the lip, and Torren hauled his legs up. He slid onto the battlement as Micca dipped down to get back out of sight. As soon as he slipped out of the way, Micca approached and, after setting his hands on the edge and getting as close to the wall as he could, folded his wings onto his back. Torren reached over and helped him up.

Keeping low, the two men scurried across the open to the other side. Staring down into the deep darkness, Torren sat still as Micca once again took hold of him. His nervousness rising, he nodded, and they dropped off the edge. He closed his eyes as the wind rushed past his face, fear tearing at what was left of his stomach.

He was jerked slightly upwards as their descent abruptly slowed. Sticking as close to the wall as he dared, Micca lowered them toward the ground. As soon as they touched down, Torren sighed with heartfelt relief.

"Are you all right?" Micca asked from behind him, sounding slightly out of breath.

"Yes," he said without turning around. "It's just been a long time."

Micca asked nothing else. After several minutes of waiting gave no indication they'd been seen, the Flyer put his disguise back on; and the two of them used the darkness to slink away.

Luck was with them, and the night's sky stayed dim, ominous clouds obscuring the moons and stars. They pushed on as far as they could until the trees closed around them, making it too dark to see.

"We'll need to stop. We can start again at first light." He was sure Micca would protest, but there was nothing more they could do.

The Flyer heaved a long sigh. "All right."

They settled down were they were.

Torren found his thoughts racing too fast to fall asleep. This couldn't really be happening. He couldn't really be going off on this fool's errand. He wondered if he would have the nightmare again, or if it would leave him now that he was on his way.

"Torren, I want to thank you again for coming with me." Micca's soft voice drifted to him through the darkness. The gratitude he heard in the Flyer's voice made him feel ill at ease.

"I've done nothing yet, and there's no guarantee I'll be able to."

"Still," Micca insisted, "I'm grateful. No matter what happens."

Torren said nothing else.

# CHAPTER 16

THE NEXT DAYS WERE A HURRIED BLUR. THE TWO MEN grabbed a few hours of sleep here and there; but for the most part, they were constantly moving.

Torren was used to walking long distances, even at this hurried pace, but Micca wasn't. With the additional restriction of having to keep his wings out of sight and his inability to fly because of it, the journey wore most on him. Horses would have speeded them along, except Micca didn't know how to ride; and Torren had no time to teach him. The pace took its toll quickly, but the Flyer never slowed or complained.

So, in amazement, he watched as Micca seemed to rejuvenate before his eyes when they first caught sight of the capital floating in the distance, even as his own heart filled with dread. It was home, a place he hadn't seen for more than sixteen years. A place he thought he'd never be returning to again. A place where he no longer belonged.

"Come on," Micca cried. "If we hurry, we can be there by tonight!"

Torren wasn't looking forward to it.

The sun had been down for hours by the time the island came overhead. Micca shed his disguise like unwanted skin, his wing tips quivering with anticipation.

"Are you sure you can carry me all the way there?" Torren asked. "I can wait here until you get some help." His gaze never left the dark mass above them.

Micca shook his head, his face shining with eagerness. "The sooner I get you there the better. Let's go."

Giving in to the inevitable, Torren only nodded.

The night was still, only the sound of Micca's flapping wings breaking the silence. As the dark mass came closer, his breathing grew more labored. Torren worked hard at trying not to let it worry him. The air grew cooler as the altitude increased.

As they neared the large mass then hovered beside it, points of light winked at them. Moonlight shone off large columns holding arched roofs. Tall spires soared into the air, their domed tops open at the sides. It stole his breath to be so close to them; his memories resonated with what he saw, reopening wounds he'd thought long-healed.

Micca reached the topside of the island, grunting with strain. Now, Torren could see a faint shimmering wall of force covering the entirety of the island. As they passed through it, he felt a light tingling sensation wherever it touched his skin. The air warmed as they emerged inside.

As if a page were turned in a book, he remembered the field was a part of the island, part of the gift. It helped keep the islands temperate so His people could travel wherever they wanted without fear. It also lessened the power of storms and rain, quieting buffeting winds to breezes and harsh downpours to gentle showers. When he'd actually faced a real storm on land for the first time, it'd driven him nearly into a panic, until his Lander hosts had explained it was normal for them to be that way.

As soon as he could, Micca set Torren on the ground. Torren turned around as the young Flyer landed and his legs gave way under him. Sweat dripped off Micca's face, his wings drooping to the sides.

"Are you all right?" He knelt, concerned about the earnest young man. A twinge of fear reminded him that, without the Flyer, he'd be alone in this hauntingly familiar place.

"Yes...fine. Just...need a moment." Micca lowered his head as if having trouble catching his breath. "I got you here."

Torren almost smiled, softly shaking his head at the Flyer's stubbornness.

"You filthy grub."

He threw himself to the side at the sound of the voice as the point of a spear penetrated the spot where he'd been kneeling. Rolling back to his feet, he stayed crouched, hand on the pommel of his sword. A guard in bronze-colored armor stood a body's length from him, a spear in his hands. The guard made as if to rush at him again.

"Stop!" Micca struggled to rise to his feet. "You misunderstand. He was trying to help me!" He moved between them. "And before you insult him any further, this man in not a Lander, but a Chosen. He is Torren, son of Lar. He was wounded doing El's will."

The guard took a stunned step back, regarding them as if he couldn't decide what to make of them. "But he's dressed as a Lander."

"And so am I," Micca spat at him. "Does that make me a Lander?" His wings shook in agitation.

The guard took another step back, his own wings drooping, yet his spear remained steady.

"I have no orders saying anyone would be coming in this evening." He sounded defensive.

Torren took off his pack and handed it to Micca. The Flyer took the hint and hunted in the side pocket for the documents sent by his uncle.

"Here—a message from Dom Rux stating the validity of my companion's identity. It's been signed with his seal."

The guard drew closer to look as Micca unrolled the parchment in the semidarkness. The embossed winged symbol of El could be seen at the bottom. He took the document and moved away so he could study it at his leisure at a safe distance.

He withdrew his spear. "Sir, I'm sorry." He rolled the parchment back up and returned it to Micca's waiting hands.

"Don't worry about it. We all do what we can for the Vassal and El, do we not?" Micca's voice still sounded strained. He glanced back at Torren, his eyes asking for forgiveness of what might have caused catastrophic consequences. He handed him back the pack then turned to face the guard. "We will need an escort to the house of Mallean the Wise. I wouldn't want to risk further misunderstandings."

Torren remembered the name as being one of the five

who'd been with Larana at the time she took ill.

The guard stared from one to the other of them, his pose submissive.

"Yes, of course. This way." He spun about and leaped into the air, heading south. He came to a sudden stop and glanced back, realizing not all of them could follow, then returned. It was hard to tell, but from what little they could see of his suddenly red neck, he was deeply embarrassed.

"Excuse me, sirs. I meant no offense."

Obviously expected to do or say something, Torren waved the apology aside. "None taken."

The guard nodded gratefully and led the way on foot. The path they followed was of stone, bordered by trees and manicured bushes. In some ways, the city resembled a giant formal garden. Here and there, statuary stood beside the paths or on pedestals. The art varied from extremely detailed realism to the surreal.

Micca sidled close, keeping his voice down so the guard wouldn't hear.

"I realize Mallean is one of the five who might be involved. But my uncle and I agree it couldn't possibly be her. She is one of the few who believes Aen was poisoned."

Torren nodded at the information but decided he would reserve judgment.

Other guards spotted them and came to see what was going on. Their escort gave quick explanations without slowing. Several of these men joined the small party—Torren felt their eyes on him when they thought he wasn't looking. He tried to ignore the sensation as best he could.

He could not help frowning, however, at the large numbers of armored men they came across. He didn't think in his youth there had been this many men going around armed after dark. The fact Landers possessed no means of reaching the islands had made it unnecessary. It looked as if they no longer believed this. Torren was sure it was one of the "changes" Rux had alluded to. He wasn't certain it was for the better.

Before long, the group reached a building fronted by columns like the ones at the embassy. Unlike the embassy, the roof sloped to form two sides of a triangle and appeared as if made of one piece. The walls were not solid but rather of lengths of blue gossamer material strung between the columns.

"Torren, please wait for me here." Micca's expression was apologetic. "It'd be best if I broke the news to her first."

"Whatever you think best." He retrieved the scrolls Rux had given him and handed them over.

The young Flyer nodded. "Thank you."

Taking the scrolls, he hurried up the steps into the confines of the house. Torren waited on the well-groomed stone path, trying to ignore his escort. He could well understand their curiosity, their confusion at seeing what looked like a Lander with the face of a Chosen. It was one of the things he'd long hoped to avoid.

He glanced up at the sky and noticed the stars and moons overhead did not stay still. Rather, they gave the illusion they were moving, though it was the island that traveled. He didn't watch the effect for long, finding it a little disconcerting. He was no longer used to seeing a sky that moved.

Micca was back after only a few minutes, followed by an older woman with a long face, piled white-gold hair and wearing a long robe. She brought a lamp with her and set it on the ground once they came near.

"Torren, I want to introduce you to Tel Mallean," Micca said. "She's been watching over the Vassal. She's one of the wisest voices in our council."

The small woman waved away the endorsements and leaned forward to get a better look at him. Torren glanced away, not feeling comfortable at the close scrutiny. He quickly looked back, however, when she took his hands in hers.

"You are your father's son—those eyes, that chin. And you've borne so much for one so young." Mallean's bright eyes bored into his.

He stared, her words disturbing him as nothing else had. He tried to retrieve his hands without seeming rude, but she wouldn't let go.

"Son of Lar, I will take you to her. It's providence you've returned to us when you have."

He looked away again, not feeling that at all.

"Come." She pulled lightly so he would follow. Rather than return to the house, she let go one hand, picked up the lamp and took a path leading toward the interior of the island. Micca and the guards followed.

Mallean's steps were short and sure, though she barely paid attention to the path. Her focus remained on Torren, one wing stretching as of to shield him. He found the whole experience unnerving. He wondered if she, too, possessed the gift of feeling what others felt through touch as Larana did. It didn't buoy his confidence.

"You have been gone a long time, son of Lar," she whispered. "You will find certain things won't be as you remember them. Your loss, and that of your father and the others—and especially the Vassal—affected many things. You must take great care in all you do."

He nodded, not knowing how to respond.

After only a short while, they reached a house not much different from the others he had seen, except a cup with wings was carved into all the columns. The curtains of the outer rooms were a startling white bordered in silver. As they reached the steps, Mallean turned to regard the others.

"You have done your duty, and we thank you," she said. "You may return to your posts."

As one, the guards gave her a half-bow, including in it Micca and Torren, then moved to go. Mallean paid them no further attention, already having turned to lead the way inside.

Like most houses of the Chosen, the *sorium*, or outer rooms, of the Vassal's possessed no permanent walls but were instead formed using drapes. Each room lead to the next, all open to each other unless a wall of fabric was erected to create greater privacy or separation. Only the central section of the home, the *lirium*, was enclosed with solid walls. It was there one could find the inner garden, or *El-at*, as well as the sleeping rooms and bathing areas.

It was to the central rooms Mallean led them now. Torren reached up and felt Larana's clip through his shirt, where he'd hung it on a string around his neck. He wasn't sure if he actually wanted to see her in her current state.

Mallean stepped through an arched doorway covered by the same white linen used for the walls. He heard her quietly whispering to someone within as Micca held the cloth aside so he could enter.

As soon as he entered the dimly lit room, Torren was flanked on either side. He'd just registered the fact they were armed and armored, his hand moving toward his sword, when Mallean's voice cracked like a whip.

"Were the two of you not listening?" Her eyes burned at the two men on either side of Torren. "This is Lar's son, a Chosen, not a Lander." She waved them off to the sides of the room, giving them another hard look as they hesitated.

"Mar, Styn, do as she says!" Micca glared at the two men as he joined them. "I left you here to protect the Vassal, not to harass the one who brought her back to us."

The two men jumped back as if struck.

"Ren Micca!" Both dropped to their knees. "Our apologies." Torren realized they were twins.

"Don't worry about it," he reassured them, "I'm getting used to it."

Mallean looked at him as if startled by his words then laughed softly.

"Are you truly the one?" A young woman with lush golden hair cascading down one shoulder stepped out of the shadows. Her soft wings were folded over her like a living cloak. The light from Mallean's lamp reflected in her pale-blue eyes. "Aen has spoken of you often."

Torren self-consciously studied the enchantress before him. Tall, graceful, she was the epitome of the beauty dreamed about by Chosen and Lander alike. She was what most would have imagined the Vassal to look like instead of the plainer Larana.

"How is she?" he asked, already sure of the answer but uncertain what else to say.

The woman looked away, her expression one of utter regret. "She...sleeps."

"Tyleen," Mallean said kindly, "take him to her."

With a slight nod to the councilor, she beckoned Torren

to follow. Leading him into the deepening shadows, she moved aside a heavy curtain dividing the room. Light streamed from within, blinding him for a moment.

Larana lay in a Lander-style bed, cocooned with blankets and pillows. He stepped forward, his gaze glued to her face.

Her expression was serene, giving no impression she was more than resting. He noticed someone, probably Tyleen, had taken the time to place her hands over her chest and comb her long hair out around her. She looked more like a fragile doll than the dying vessel of a god.

Anger ignited inside him. Why would someone do this to her? Until mere days ago, Larana hadn't even heard of the Chosen. She was a carefree soul, a young, bumbling girl—and deserved better than this. Torren was surprised as his vision suddenly blurred, his throat growing tight. With the greatest care, he reached for one of her hands and held it.

Nothing. This time he felt nothing. Every time before when he'd touched her exposed skin he'd felt something—whether fear, grief, worry, joy—but never nothing. He put her hand back, his own shaking. He swore he'd do what he could to help her.

Taking one long, last look at her, Torren turned away. He hesitated, having forgotten Tyleen was with him, as he found her watching him intently.

"Please take good care of her."

The Flyer looked away. "Leave it to me. I will protect her with my life."

Metal glinted as she revealed a small, thin blade hidden in her robe. He nodded, trying not to let his shock show, and left the curtained room.

Micca waited anxiously, his face strained. "How does she look?"

"Well. Without pain."

The Flyer nodded slowly.

"Why don't you go see her?" Mallean suggested.

Micca refused. "I don't deserve to. I failed the Vassal once, and until I make amends and she is once more awake, laughing..." His eyes grew dark.

Mallaen placed her hand on his arm. "El will not desert us. The Vassal will be saved."

His wings rose, and he glanced at the councilor gratefully.

"You have a few hours before dawn. Both of you look exhausted, and tomorrow may prove even more taxing," she stated. "You can both stay here, in the guestroom. Already rumors of your arrival will be spreading, but no one will dare disturb you in this house. I will return in the morning, and we can discuss how to proceed from here."

Micca opened his mouth to argue, but a raised brow from Mallean made him keep whatever objections he had to himself.

"Come with me." Not waiting for them to acknowledge her, she turned and exited the room. Following the outer wall of the inner chambers, she stopped before the next open doorway. She slipped inside, glancing behind her to make sure they were following.

By the time Micca and Torren joined her, she had taken a taper and lit another lamp from her own. The room was long and spacious, the ceiling high. A starry landscape had been painted on the ceiling, the three moons and their surrounding auras shaped like women's faces. Two of the Flyer-style beds sat at opposite ends, folded blankets at the end of each.

"The bathing and other facilities are through there." Mallean pointed toward an opening on the other end of the room. "I'll see the two of you again in the morning." She smiled at them and left.

Micca sighed and sat down on the farther of the beds. "I'd half-hoped she'd wake once you went to see her."

His voice was filled with longing.

"It would have made things easier." Torren admitted, staring at the other Flyer bed. It appeared he would be sleeping on the floor tonight.

Micca settled back, his wings falling naturally to either side of the thin upper half. "Oh, how I've missed this!"

Torren glanced at him, then set down his pack and retrieved a change of clothes. "Will you be needing the light?"

When he got no response, he looked toward the young

Flyer and found Micca's eyes closed. He had already fallen asleep.

Not having any intention of indulging in that quite yet, Torren took one of the blankets and covered him with it before retrieving the lamp and heading to the next room.

A partially raised tub dominated the floor of the bathing area. A bucket full of fuel sat next to a grill opening on the north side of it. The stone tubs were lined with metal to disperse the heat of the small stove to the water more quickly. It was something Landers had yet to figure out how to do.

To the Chosen, bathing was almost a vocation. Large water storage tanks were kept on all the islands and were regularly filled with either water gathered from the rains or collected and brought back on the flying ships. Complex piping systems carried the water to all of their homes. This was also something Landers had not yet perfected, though in the cliff cities of the south they were ahead of most. And the majority of the other large cities had bathhouses.

Still, elsewhere bathing was not as important. Unless a river or other water source was nearby, dragging and heating water was a chore not all could afford to spend the time on. One well-known Chosen complaint had to do with the ever-pervasive Lander stink.

Torren stripped, not bothering to light the fire, not wanting to wait for the water to warm. He needed to dispense with this now when no one was watching—there were still things he had no desire for anyone to see.

After a cold bath, which left his skin covered in goose bumps, he redressed and returned to the room. The Flyer hadn't moved, as oblivious as before. Taking the blankets from his bed, he arranged them on the hard stone floor and settled down, putting out the lamp. He lay in the darkness, his hand wrapped about the hair clip still tied around his neck.

# CHAPTER 17

TORREN DREAMED, BUT IT WASN'T THE OLD NIGHTMARE or even the one that had kept him in Caeldanage waiting for this unwanted journey. In this dream, he saw himself as he was now. He stood before a large plain, his arms extended to either side. Pain twisted his features as the view slowly slid back, and he could see he was being pulled from both sides and nothing he did would set him free.

The image continued to pull back until he saw the hands that were holding on to him. On the one side were weathered hands, the hands of Lander farmers, their wives, city guards, barkeeps and merchants. The others belonged to young and old Flyers, some wearing the short drapes, others in long robes—councilors, ambassadors and artists. Each side wanted him; neither would give him up. And, excruciatingly, they were tearing him in two.

A soft touch on his shoulder brought Torren awake with a gasp.

"Oh!" Tyleen pulled back, startled as he sat up struggling for breath. "I'm sorry, I didn't meant to alarm you."

As she apologized, she stood up and got out of his way. He grunted and got to his feet.

"My shift was over, so Mallean asked if I'd wake you before I left. She's waiting for you to share the morning meal." Tyleen spoke quietly, appearing wounded as he still said nothing. Her wing tips quivered slightly. "Micca has already gone to the bath. I...brought clean clothing for the both of you." She indicated some folded garments on a table standing against the wall.

Torren looked at her for the first time. "Thank you for the offer, but what I'm wearing will suffice."

Tyleen's sudden disbelief made her bold. "But, would you not...? I had thought..." She glanced at his Lander clothing, her face filled with distress, then looked away.

"I'm perfectly comfortable as I am." Slivers of his fading dream poked at him, but he knew it wasn't the real reason for his decision. Whether Tyleen realized it or not, he'd be a total spectacle if he tried to dress like them. It would expose things he didn't want seen. Nor would he explain himself.

"As...As you wish," Tyleen said presently. "I will return to take you to Mallean presently."

Without looking at him, she rushed out of the room. Holding back a sigh, he picked up his blankets and tried to think of nothing but the task of folding them. As he set them on the bed, he couldn't help but feel coming here might have been a monumental mistake.

"Torren! Good morning." Micca appeared, water shining in his hair. The Lander clothes he'd previously worn had been discarded and replaced by a light green Flyer drape.

"Morning."

Micca's brow arched at the lack of warmth in his voice and then went even higher when he noticed his Lander clothes. "Did Tyleen not tell you about the new clothes?"

"She did."

His roommate stopped and stared at him for a moment, then nodded and sat down on the bed. He grabbed a pair of sandals and put them on. "I take it you don't find Lander clothing restricting, then? I swear those leg coverings chafed me in I don't know how many places."

He offered a smile.

Torren shrugged. "You get used to it." He turned away, hoping Micca would drop the subject.

"I don't mean any offense," the Flyer went on, "but the others will find it more difficult to think of you as one of us if you're wearing Lander clothes."

Torren didn't look at him, trying to cover his growing annoyance.

"Even if I changed, they'd still take me for a Lander." His tone was harsh.

"Perhaps, but at least then they would see..." Micca came to a stop as Torren sent him a scathing look. "Ah, perhaps you're right. I'm sorry."

An awkward silence settled between them.

Torren ran his hand through his short hair, exhaling heavily, letting his anger go. "If you're ready, we should go. I believe Mallean is waiting for us."

Almost as if his words summoned her, Tyleen peeked in through the doorway. She avoided looking at Torren, but on spotting Micca brightened. "Are you ready?"

He stood, straightening his short drape. "Yes, please show us the way."

She led them to the outer eastern section of the sorium. The screening curtains were pulled back, allowing in the morning breeze and bright sunlight. A table had been set close to the edge with several chairs. Upon their arrival, Mallean stood and welcomed them. After bidding them to sit down, Tyleen excused herself and left. Mallean made no comment on Torren's style of dress.

"Please, eat your fill." She gestured at the plentiful spread before them.

Suddenly ravenous, Torren filled his plate with sliced ti eggs from the eastern swamp reaches, eva fruit from the west, golden apples from the north and other dishes he could not readily identify. As he ate, the flavor of some of the unknown items triggered his memory, reminding him he'd eaten them in his youth, though he was still hard-pressed to remember what they were.

With the Twenty Islands of the Chosen floating around the world and trading amongst themselves and specific Landers, goods from all over were at their disposal. Granite for their buildings, wood from mountain forests, meat from herds kept on mesas where no Lander could reach.

"I hope you were able to get some rest?" Mallean said as their hunger finally began to abate.

"Yes," Micca answered with feeling. "It felt good to be back home."

Mallean nodded, half-smiling. She then gave Torren a questioning look as he said nothing. When he continued with his silence, she didn't press him.

"The letters you brought with you have been delivered. News of your arrival has also spread with the morning light. Requests have already come for Torren to appear before the assembly. The return of two who'd been thought lost to us has everyone very eager to see you."

He was unhappy at the information but not totally surprised. "When?"

"As soon as we're through here, if it would be convenient."

"Is an assembly such a good idea?" This came from Micca. "I'd hoped our presence here would be less publicized. If some of the matters are as we fear..."

Mallean nodded. "It can't be helped. As I've said, Torren's arrival has already been announced to every household on the capital. It would seem the two of you made quite an impression last night."

Micca gave a small grimace.

"And perhaps," she added, "this way, some who might not have will now cooperate, since you'll be more exposed to the public eye."

He didn't look convinced but didn't argue.

"Torren, I must caution you not to make mention of the Vassal's current state."

He looked up at Mallean's sudden quiet tone.

"While the truth is known by all in the council and a few others, everyone else believes Aen has only taken slightly ill due to the strenuous ordeal that brought her back to us." The councilwoman suddenly looked old. She stared hard at her hands. "It is a thin lie, but one most of our people prefer to believe rather than the reality."

"I understand." He remembered the panic and fear that had gripped the Chosen when the Vassal first disappeared. How much worse would it be, after fifteen years of waiting, to lose her again after only a few days? Unless they were able to discover what had been done to Larana and reverse it, he might get to find out.

The three sat in silence for a time, what little appetite any of them might have remaining gone and forgotten. Torren felt his insides tightening.

"Could we go get this over with?" He found the suggestion coming out with more distaste than he'd meant it to.

The other two nodded, taking no offense. They all stood up to go, Mallean studying Torren from the corner of her eye.

"This way."

She led them out of the Vassal's home, her sandals making soft slapping sounds against the granite sidewalk. Torren looked up as they followed a bush-lined path, the sound of flapping wings overhead startling him. Giggles and frightened ahs floated down to them as wide-eyed children swooped to stare at him before fluttering away. A guard on a spire glowered at the fleeing youngsters but then turned to goggle at them as well.

Torren spotted other Chosen flying in their same direction, the clouds moving at a faster pace than he was used to as the island continued its travels. The open amphitheater he recalled from his youth rose before them. His apprehension grew.

Atop each of the tall spires along the way he had spotted one or two guards standing at attention, their bronze or silver armor gleaming in the sunlight. Once more, he was troubled by the increased militant atmosphere among the Chosen. Some things had definitely changed since he'd last been here.

Walking on either side, Micca and Mallean escorted him into the amphitheater. Though all of those he saw took to the air to go inside, Mallean led him to a wide opening on the structure's south side. As the bulk of the building rose before him, Torren felt a pressure growing inside.

As if knowing his feelings, Mallean linked her arm with his, a smile on her face. "This will mean so much to them."

He could only nod, not trusting himself to speak.

As they proceeded down the dark hallway, a buzz of excited conversation drifted toward them. Mallean stopped as they reached the end, leaving them in shadow and out of the immediate view of those outside.

"Wait here, please." Squeezing his arm in reassurance, she let him go and stepped forward alone into the open arena.

The voices rose to thunderous proportions as she appeared, all faces turning eagerly toward the councilor. The front rows were colored blue and were filled with the representatives from the Twenty Islands. Behind them sat the citizens of the capital. They grew silent as Mallean reached the center of the floor.

"Good people!" Her voice rang out, amplified by the building's acoustics. "Today we have cause to celebrate. One who was lost to us more than fifteen years ago has returned. He is the sole survivor of those who did not come home after they valiantly went in search of the Vassal. Please welcome him back into our midst, and let's rejoice at his return."

Mallean half-turned and held her hand out in a welcoming gesture. "Welcome home, Torren, son of Lar."

His legs froze now that the time had come. He stared at the sea of faces eagerly awaiting him, and it frightened and angered him at once. He was alive, but only by chance. He was no one; he could offer them nothing. Those who would be worthy of a homecoming, unlike himself, lay long dead in their graves. These people would want things from him he couldn't give.

"Torren, it's all right." Micca stood beside him, his face shining with confidence. "They're waiting for you."

"Torren! Torren! Torren!" More and more of the Chosen rose to their feet, calling his name. Mallean remained facing in his direction, waiting for him.

He finally forced himself to move, feeling nothing but icy dread as he stepped out alone into the light. Hundreds of eyes locked on him, some voices growing silent as they saw him for the first time. He felt his insides churn, only having felt this exposed once before in his life.

Never had he meant to see another Chosen. Now, he was before a city of them, all staring at him like he was some oddity in a cheap carnival. Thousands of eyes like tiny prickling needles, trying to see what lay beneath his skin.

Murmurs ran back and forth as he neared Mallean, some in Common but others in the language of the Chosen. He wasn't sure they were all favorable. Still, some cheered as he came to stand at her side.

"Tell us what happened!" This came from the vicinity of the councilors' seats. Like a litany, others picked it up until it rang as loudly as those still shouting his name.

"Won't you tell us?" Mallean asked him in a soft whisper, laying her hand gently on his arm.

Torren bowed to the inevitable and slowly nodded. Mallean held up her arms for silence. As the assembly grew quiet, he told himself over and over again none of this mattered. Looking at no one, yet feeling Mallean's buoying presence beside him, he retold his story.

Whispers trickled back to him as others passed his words to those who couldn't hear or translated for those who didn't know Common. By the time he was done, Torren wanted nothing more than to be gone from there.

"Thank you," Mallean whispered to him. "It'll be over soon." She turned to face the crowd. "There is one more thing you must be told about the son of Lar." She paused; and Torren felt his stomach drop, a cold feeling telling him he knew what was coming next. "He was the one responsible for returning Aen to us."

Shocked gasps peppered the crowd. His jaw tightened. A cheer grew, resembling a wave, and spread until it thundered around him. El's name as well as the words *destiny, mercy, honor* passed through the crowd. Soon, the whole of the audience was on its feet, shouting his name again. The amphitheater reverberated with the sound.

He couldn't bring himself to look at them. He had no right to this. Luck was responsible for all of it—chance, not any true intentions on his part. If they only had an inkling of his thoughts on El and the Vassal over the years, they wouldn't cheer him but condemn him. It should have been his father standing here, his father's name the one ringing from their lips—anyone's name but his.

"Good people!"

Torren looked up as he recognized the thundering voice. It was one of the three councilors he'd met at the embassy—Councilor Valerian. The middle-aged man flew from his seat to land in the amphitheater's open circle. His charismatic face and demeanor commanded the attention of the crowd.

"You see before you proof of our Lord's power. Mysteries left buried have now been revealed after so many years of waiting—but there is still more we do not know, more that needs to be done. We cannot forget there are issues still unresolved. So, your elected council must convene and speak to our lost one.

"Go home and celebrate these bright tidings. Later, when we are done, welcome our lost brother back into the fold, but for now, the council must tend to El's business."

Groans of disappointment filtered throughout the crowd. Whispers and muttered questions drifted down to the floor.

"Whatever we discuss or find out will be revealed to you," Valerian admonished. "We will give him back to you very soon!"

Though still unhappy, the people started to depart, taking flight in ones and twos. Within minutes, most of the stands were empty.

Valerian turned to face Torren and Mallean and strolled over to where they stood.

"The council has decided to meet indoors, away from the curious," he said. His hard eyes raked Torren. "This way."

Micca caught up to the trio as Valerian turned to show the way. The other councilors left their seats to join them, heading in the same direction. All but one.

A woman in her later years stood facing them, tears trailing down her face. Torren frowned as he spotted her, something about her appearing familiar. The woman was reserved, her robe simple, yet it bore the stripe on the sleeve proclaiming her a councilor.

"Do you remember her?" Mallean's eyes were bright as she asked, having already noted the subject of his scrutiny. "She's been wanting to see you, but waited until it was proven you'd truly returned."

He continued to study the woman as they drew closer, the conviction he did, indeed, know her growing within him, the soft dark-blue eyes, the way she carried herself.

Time and grief hadn't been good to her. Lines surrounded her eyes and furrowed her face. Her golden hair had turned white and lost a lot of its lushness. Torren felt his chest tighten. He'd told himself for years he wouldn't ever see her again. What would she make of him now that he was no longer whole?

"Ze–Zelene?"

At the sound of his voice the grave woman's face split into a brilliant smile, as if all doubt had been erased with the one word.

"Torren, Torren, you've truly been returned to me!" She drew him into a tight embrace. "My son."

For a moment, he thought she'd taken lessons from Sal, as he found it hard to breathe, but the thought was quickly thrust aside as she almost collapsed in his arms.

"Mother!"

He reached out for her as she sagged and steadied her on her feet. It hurt to see her, to see what time and sorrow had done to her; but at the same time, it felt wonderful.

Micca and Mallean crowded close in case he needed a hand. Zelene looked up, joy infusing her eyes and face.

"You came back, even though they told me you wouldn't. You came back to me. El be praised!"

He looked guiltily away, knowing if matters hadn't turned out as they were he wouldn't have come at all. It was then he noticed how Micca and Mallean had spread their wings protectively about them. They were trying to give them what privacy they could as a number of the other councilors stopped to stare at the reunion.

"Maybe we should take her home," Micca suggested softly. "I don't think anyone in the council would object."

Zelene stared up into her son's face, as if wanting nothing more than to be alone with him, but then pulled away, wiping her face with the edge of her sleeve.

"No, no, it would be selfish of me. El has returned my son to me, and now I must help with El's business. We will have time after."

Her eyes sought his, and Torren nodded slowly.

"Will you do me the honor of sitting with me?" she asked.

"Of course."

Zelene clung to her son's arm as they started once more on their way.

Outside the north exit of the amphitheater stood a

shorter building with solid walls behind the tapered columns. As soon as all the councilors entered, thick redwood doors closed, sealing them in.

Every island had four representatives, each wearing robes with golden wings embroidered on the right shoulder and a purple band on the sleeve. Interspersed amongst them were a few family members filling the role of aide.

The room had padded benches set around a small open area in the center. The design, similar to that of the amphitheater, also resembled the temples of El.

Torren sat down next to his mother, still trying to reconcile himself with how much she'd changed. She, on the other hand, beamed at him, cradling his hand in hers as if afraid he would disappear. Mallean and Micca took the bench directly behind them.

During open meetings, his father had often brought him along so he could watch how the body charged with carrying out El's will worked. His gaze drifted around the room, spotting some faces that seemed slightly familiar, though most he didn't recognize at all.

Looking for them, he found those who might have possibly had something to do with Larana's current state. Icos seemed even more bowed than before. Mides didn't appear much better. In truth, most of those gathered here appeared somber and weighed down by the knowledge they were so determinedly keeping from everyone else.

Valerian he'd already seen; and since he'd arrived at the capital, he'd met Mallean. The only councilor possibly involved he'd yet to meet was Symeas.

As if reading his mind, Micca tapped his shoulder and leaned forward to whisper in his ear.

"Symeas is the middle-aged man sitting four places from Valerian in the back."

He was a mouse of a man, talking quietly but animatedly with the councilor sitting next to him.

As everyone settled into seats, Valerian rose and stepped out onto the central floor. A thin silver stripe within the purple heralded him as the arbitrator for the council. This was a position that, up until her disappearance, had traditionally been held only by the Vassal.

"Fellow councilors, it is time for this meeting to begin." His voice boomed through the room. He waited a moment for them to quiet down. "All the knowledge gleaned by those selected to confirm the Vassal's identity has now been imparted to you from Lar's son's own lips. And while we're grateful to know what befell those who were taken from us, it is a pity not enough is known to point us in the direction of those who instigated the tragedy.

"Worse, we do not even know if those involved had anything to do with our more current troubles."

Torren frowned, wondering if the tone of condemnation lacing Valerian's words was aimed at him or if he was just imagining it. Murmurs circulated amongst the councilors.

"Presently, we are less than two days away from Caeldanage," Valerian continued. "The Vassal's state has not degenerated, but neither has she improved."

"She will!"

"She must!"

Valerian tolerated these outbursts, his expression clearly showing his doubts on the matter. "We will find out soon enough. But..." He held up a hand to forestall further interruptions. "If the unthinkable happens and Aen does not awaken, we must decide now how the people will be told."

Several councilors stood up throughout the room. "We can't tell them!" "It would be chaos." "They'd want blood." "Suicides!" "Madness."

Others stood up as well. "But we must!" "They have the right to know."

"Everyone, please!" Valerian's booming voice overrode them all. "One at a time."

The shouts subsided, though many of the councilors remained standing. The nervous shuffling of wings whispered across the room.

"Tel Lec, if you please."

The particular councilor's face was blotched with barely restrained emotion. "We cannot...We cannot tell them. It would destroy our people." He stared earnestly at his colleagues. "The black pall that has hung over us for so long they all believe has been lifted. If they were to find out we've been cursed yet again..."

"Hah!" Icos rose, smacking his cane against the floor.

Rather than picking one of the others already standing, Valerian deferred the right to speak to him. "Fia Tel Icos."

"They must be told!" The old councilor smacked the cane against the floor again, making a resounding thump. "This is a test, a test of faith. It is our payment for having taken so long to find our Lord's Vassal. Only once everyone has been told of what's happened can we as a people pray for forgiveness and have our God's link to His people returned to us again."

He thumped his cane one last time and sat down.

"Tel Arlean." Valerian indicated someone behind where Torren and the others were seated.

A woman's high-pitched voice resonated across the room. "What none of you seem to understand is that it doesn't matter whether the people are told or not. El discarded the child as his Vassal as soon as she was taken by the Landers."

Cries of angry protest resounded across the room. Arlean ignored them, raising her voice to be heard over the din.

"You're all aware she cannot speak our tongue. You all know she barely has any knowledge of who El is or why the Vassal is so important."

"But she has the gift!"

Arlean ignored the comment. "El hasn't spoken to her, He's never guided her. And once she was returned to us, what happened? She collapsed, fell into a sleep no one can wake her from. It is El's doing! He's telling us He doesn't want her—she is unclean!"

Mallean shot to her feet at this, her eyes blazing. "And what would you have us do, Arlean, kill her?"

The room grew suddenly quiet. Arlean's thin mouth worked but no sound came out. She abruptly sat down.

"If it's true, she must be purified." This came from across the room.

"How? El is not here to guide us."

"We make our own way. We decide how it is to be done."

A full furor followed this last pronouncement. Voices rose, feathers separated from agitated wings. Some started to yell at one another.

"Enough!" Valerian shouted at them, his face hard. "This squabbling will fix nothing." The room settled down, harsh looks flashing between some of those present. "The question must be decided, and we must be united on the reasons we will give the people. We can't have them going at each other's throats over unfounded opinions!"

"What would unite us all is going after the Landers and making them pay."

Torren, as well as most of those present, turned to look at the young man who'd spoken. He seemed about Torren's age; and though he was in the room, he didn't wear the robes of a councilor.

Shouts of agreement and others of dissent peppered the room, mixed in with a few horrified stares.

Mallean rose to her feet again. "Elon, it is not your place to speak. You're here only in order to observe the meetings for your father."

The longhaired young man stared back at the councilor haughtily. "I have only said what's needed to be said. It's what the people will want. And if my father weren't so ill, he would be here telling you this himself."

"You don't know your father, then." This remark was followed by a number of unkind chuckles. The mood grew darker in the room.

A heavyset councilor rose to his feet. "I, for one, believe we should talk about this, whether or not young Elon had the right to speak." He surveyed the assembly. "Though we haven't done anything against them, the Landers stole the Vassal, the Landers killed our people and now the Landers have placed Aen into an endless sleep." He paused for a moment, letting his words sink in. "They are the cause of all our misfortunes. They are the ones who've taken El's grace from us."

Mallean spoke again. "But we have no proof it was the Landers, and even if it was, which ones?"

"What does that matter?" cried another. "A Lander is a Lander is a Lander. If a Chosen had done something to one of them, do you think they'd care which one of us it was?"

Symeas stood up in indignation. "Are you now comparing us to grubs?"

"No, I'm just saying they're animals, uncivilized, and should be treated as what they are."

"What about El's tenets, our history? We came from Landers." A handsome woman of thirty or so spoke from across the way.

"Yes, but that was a long time ago. They've degraded since then, lost their intelligence. They're ugly, violent, short-tempered. You know what they did to our ancestors when all they tried to do was help their sick and needy. We were repaid with pain, disgust and abuse. They're no better than beasts."

"There's no need to insult the animals!"

A couple of snickers answered the gibe.

Torren was rising to his feet before he realized he was going to do it. His mother and those around him stared at him in amazement, though no one was as surprised as he was.

"I am not part of the council, but I have lived amongst the Landers and know something of their ways. May I speak?"

The room grew quiet; all eyes turned to Valerian. The council leader stared at him raptly, as if weighing what he might say. After several moments, he nodded.

Torren turned to face them. "Before the time of my father's death, I hadn't had much exposure to Landers or their ways. Like all of you, I grew up with the stories and fables all parents tell their children about them. Who here during their childhood didn't, at one time or another, get told by his parents he would be taken by them for being bad?"

This elicited a number of grudging chortles.

"Over the years, I've learned some of the things we think of them are true. Some of their number are greedy, selfish, war-mongering. But I've also learned these things do not apply to all of them.

"As a people, the Chosen are united. What affects one, affects all. The Chosen have been unified in peace since the beginning, but for the Landers it's different. They have no cause or gifts to bind them together. Instead of one people, the Landers are many. Instead of one god, they have nine—some of them more. Some are together because of convenience, others due to the strength of one man or a small group, while a few are together because of an ideal.

"Each group is in many ways the same as the others, and in other ways not at all. Each strives to keep themselves safe from the others, or to take over their neighbors before they themselves get taken over."

He paused a moment, realizing he held all of their attention.

"For years—for *years*—I wanted nothing more than to kill every last one of them. I wanted to kill them for the kidnapping of the Vassal, for the death of my father, for what was done to me. I no longer feel this way. I've come to know them. I've come to understand the things I've just told you.

"Moreover, by learning their ways, I can say that whoever did these things is not allied with one of the Lander governments. Landers can't reach this island. Landers can't fly. And until they find a solution to that problem, they wouldn't dare antagonize the Chosen in this way. It would gain them nothing, and lose them much. It would hurt their trade, their economy, their cash flow—and no Lander empire can survive without money. They would have no reason to do any of this."

"Torren..." Micca's warning whisper hissed across to him. He was on dangerous ground, and he knew it. Yet something had needed to be said. The Chosen weren't a warlike people. Though he himself had done it, he had no wish to watch them soil themselves that way. Once they started on this path, they would never be the same again.

He sat down. No one said anything for several heartbeats, until Elon erupted from his seat.

"What drivel is this? Of course, the Landers did all this! They want to hurt us whether there's an advantage or not. They're jealous of us, of what we have, of our God, of our way of life! They'd do anything to see it destroyed. Perhaps you'd know this if you hadn't been turned to their ways. And of course," he continued, his face livid with anger, "that's assuming you're really even one of us—as I, for one, have seen no proof of yet!"

Torren stared impassively at the red-faced man, not

caring one bit what he did or didn't think of him. So, he was quite amazed when his mother rose to her feet, her hands bunched into fists at her side.

"How *dare* you?" she demanded. "You're not even a councilor, no more than a fledgling, yet you would question whether this is my son when I and others more knowledgeable than you have said he is without a doubt?"

"What else could I assume?" Elon countered. "Look at him, walking around flaunting his Lander clothes, his knowledge of Lander ways. What else could I think? Have him prove he is who he says. The answer might just surprise you."

"Zelene." Mallean reached for her as his mother shook with indignation.

"This is my son. He is a Chosen!" Her hand rose to point accusingly at the young upstart. "You do not belong here."

Elon's face twisted in rage; but before he could reply, Valerian stepped into the line of fire.

"Enough of this! It's become obvious nothing is going to be resolved today. This session is over. We shall reconvene on the morrow." His gaze swept the room. "Don't forget, what's been discussed here is to remain here. Today is a day for celebration as far as the people are concerned, and they will be expecting you to join in the revelry when you leave here. This meeting is adjourned."

With this, he made his way to the doors and swung them wide before stepping out into the light.

# CHAPTER 18

ZELENE TURNED TO TORREN, A SUPPLICATING LOOK ON her face. "Please, ignore what the young fool said. All who matter here know who you are."

He avoided her eyes, more disturbed by what the words had done to her than what they'd ever mean to him.

"It doesn't matter. I don't care what he thinks."

With a grateful look, she squeezed his arm.

Instead of departing, a number of the councilors pressed close to get a better look at him and to congratulate Zelene on her son's return. Micca and Mallean tried to fend off as many of the questions as possible while slowly working the pair outside.

Once there, however, things only got worse. Now that the general populace had discovered the council meeting was over, they flew in from all over the island to get their own close look at the returned wanderer. Many of the councilors put on fake smiles as they made their departure, but Zelene's was genuine and bright. With each congratulations, each blessing, it seemed to grow more radiant.

For his part, though, if not for the fact it made her seem more alive, more like the woman he remembered, Torren would have enjoyed nothing better than to shrink away and hide.

The well-wishers thrust gifts at them—hastily prepared food baskets, little trinkets, clothes. Old people told him they remembered him—how energetic he'd been as a child, how handsome, always at his father's side wherever he went. Others closer to his own age shyly revealed they'd been his playmates. A few faces and voices here or there did seem familiar, but he didn't really remember any of them; he'd spent too many years trying to forget.

"Who rescued you?" someone shouted from the back.

"You were lucky you didn't die of infection," piped another. "Isn't it true they only bathe once a year?"

Suddenly, it was as if everyone wanted to know. The crowd pressed in even closer.

"No, that's no true," Torren replied over the growing din. "They—"

"Did they make you join a mating party?"

"A what?" he spluttered. Where were they getting this?

"I heard they consider Chosen flesh a delicacy," a woman said off to the side. "Have you ever had to fight anyone off? Did they save you thinking to have you for dinner?"

"No! That's insane. Why would you ever...?" Torren felt his frustration starting to well toward anger.

Sensing his growing distress, Zelene and Micca tried to hold some of the inquisitors back as Mallean stepped in front of him and raised her voice.

"Please, he's only just come back to us. He can't answer all your questions in a day. Let's let him rest, be with his family. Sooner or later, all your questions will be satisfied."

Torren breathed a sigh of relief as the three then led him away.

Yet it wasn't the questions that really wore him down. That arose from something totally unexpected.

It was the Chosen themselves, the way they looked. All around him was a sea of light-skinned faces, blond to white hair, blue and green eyes. The majority of them were lithe, thin and perfect. It was as if he'd been taken and, with only slight alterations, replicated hundreds of times. It was nothing like the sea of variety that characterized the Landers, where, from one person to the next, looks and coloring varied. It was daunting, as if he were losing himself in a sea of Chosen. It made him feel ill.

It was hours before they were able to reach Zelene's home. He had only felt this exhausted before after a prolonged battle. Repeatedly, he'd been asked to explain how he'd found the Vassal, to tell them what had happened to him. It quickly became obvious most of the details of Aen's return had been kept within the council, for it didn't appear his part in it had been revealed to any others until today.

Torren glanced at his mother, struck once more by the unfathomable reality that he was with her again. Ashamed, he realized that, in all the years of his self-imposed exile, he'd not once considered how the lack of knowledge of what had happened to him and the others would affect her. The grief. The wounds, now freshly re-opened. And the new ones created when she'd learned he was alive but had elected not to return to her.

He found a chair and sat down, slumped forward, arms on his knees. Mallean and Micca released the outer curtains to hide them from view, letting it be known they sought privacy. Pushing his hair back with his hand, Torren found the object of his musings studying him from across the way. The moment their eyes met, Zelene smiled. He tried his best to return it, though inside he was filled with nothing but trepidation.

"Zelene, is that you?" A woman closely resembling the councilor stepped out from the lirium.

"Lii! It was true, it was true." Zelene rushed forward to meet her then pointed at Torren. "Look!"

A quick smile lit the woman's face as she turned his way. "I prayed it'd be truly you," she told him. "Zelene has always believed you were out there somewhere. I couldn't see clearly from my seat, but I'd hoped so much. I'm so glad. Welcome home, nephew!" Lii gave him a light hug where he sat. "Do you remember me?"

Torren nodded. He hadn't remembered her immediately, but the mention of her name had done it. Lii was his mother's older sister.

"Yes, I remember you. You're my Aunt Lii."

"Oh, Zelene!" Lii smiled at her sister, who beamed back. "It truly is a day for celebration." For the first time, she noticed the others. "Oh, how are you, Tel Mallean? And who's this young man?"

"I'm fine, Lii. You're looking well." She gave Torren's aunt a small, tired smile then nodded toward her companion. "This is Ren Micca, a friend of Torren's."

"Ma'am." The strain of the day was showing on his face as well.

"You two will stay with us for dinner, won't you? We can make this a big reunion celebration. I'll even cook my famous valmion tarts." She glanced at Torren. "You used to love those whenever you came to visit." Her wings rose and fell in excitement.

"Thank you for the invitation, but I think we'd only be intruding."

Zelene opened her mouth to protest.

"Besides," Mallean continued, "Micca and I have some pressing business to take care of." She turned to Torren. "Enjoy your reunion. We can pick up on things again on the morrow."

Torren felt his pulse rise as he realized he was about to be left here alone with his past. The departing visitors gave all three of them a half-bow and took their leave.

"Well, if I'm going to get the feast prepared, I'd better start now." Lii grinned at them. "It's been a while since I've gotten a chance to cook a fancy meal. And before you ask, Zelene, no, you can't help. You spend time with your son, wallow in the gift El has given you. I'll come for you both when it's ready."

With light steps, she left them.

Torren and his mother were now alone. Rather than face her, he surveyed his surroundings. The curtains were dark green with splashes of gold. He thought before they might have been sky blue. Murals interspersed with tile-work covered the outer walls of the lirium, several of them seeming familiar. Large potted plants in various shapes and sizes dotted the outer rooms.

Though he'd spent his childhood here, the very familiarity made him uncomfortable. He was in no way eased as he felt his mother's eyes on him again. He had no idea what he should say or do.

"Despite the troubles in your life, you've turned into a well-formed young man." He turned to find her admiring him. With a tentative hand, she reached to touch his close-cropped white-blond hair. "It's so short. Is this how Landers prefer it?"

He fought not to shiver at her touch. "Some do. I just find it more convenient." Most Chosen kept their hair long, past their ears or down to their shoulders. It was yet another way he'd differentiated himself from them, though a lot of it truly had to do with ease of care, and not having it get in the way during combat.

Zelene studied his face, tears glinting in her eyes. "Did they treat you well? These Landers who found you and helped you?"

Torren looked away. "Yes." As he forced himself to confront her expectant face, he realized he owed her more than his one-word answer. If anyone should know about his life, it was she. Telling her would maybe make up a little for when he left her again.

"Their names were Alain and Darrek. They were on their way to market when they ran across me on the road..."

Memories rose unbidden of awakening beneath the dirt. His horror as he gasped for breath, struggled to dig himself out, feeling the bodies of the others beneath him, the stench of their already rotting corpses invading his nostrils.

The world above had been dark, strange and much like the story of El he'd been told as a child. He'd run, driven by the fear the men would be back, pain and madness at his heels. He'd run and prayed and prayed until he knew El would not save him, until he knew there were no gods, only fear and anguish. He'd run until the blood loss drove him down and he didn't move again.

"They thought I was dead, at first, but when they checked on me, they realized I wasn't. Alain wrapped me in her shawl, and her husband turned the wagon around and took me to their home. They told me later they knew immediately what I was. The clothes, the shoes and the wounds on my back told them."

Alain had regaled him with the story often, always with a look of wonder on her face. Barren herself, the wounded boy had seemed to her a gift from the gods. It hadn't mattered he wasn't of her own kind, though it did to him.

"They cleaned my wounds and cut off the jagged edges

as best they could. Luckily I was unconscious through all this, with a raging fever."

Tears rose in Zelene's eyes and silently made their way down her face. Maybe this hadn't been such a good idea. But now that he'd started, he was finding it difficult to stop.

"They cared for me as best they could, getting me to drink a little and keeping my wounds clean. I woke up after a couple of days." He wouldn't tell her of the terror he'd felt as he realized he was trapped in a Lander home, how all the tales he'd heard about them had crowded around him, insisting they were true despite all his father had tried to teach him. He would be eaten, he would be flayed alive, his eyes placed in jars until they rotted away.

He'd screamed when one of them tried to touch him, too terrified to understand Alain's attempt to soothe him. He bit Darrek when the latter tried to keep him from leaving the bed. He only stopped because his body gave out on him and he'd not been able to try it again.

He'd understood a word here and there as his rescuers conferred in whispers, making sure to block the the door. They finally decided to leave him be for a while. After they were gone, he tried to get off the bed again. He was able to make it to the floor, but that was as far as he got.

When he'd awakened later, he was once more in the bed, warm covers around him. Alain sat nearby on a chair, watching him. How dare they keep him a prisoner? He had glared at her with all the defiance he could muster, mostly because it helped keep back the fear. Why were they waiting so long to eat him? Why didn't they just kill him? Then he would be able to join his father and the others.

But it was not to be.

"They taught me their language while I healed, adding to what I'd already learned of it here."

Though he had only listened, never speaking, wanting the knowledge for the edge it would give him but not wanting them to know he was learning. Every kindness, every deed, was met with stubborn resistance. Only when his nightmares wracked him would he show himself vulnerable and cling to Alain until sleep took him again.

They should have left him where they'd found him. They should have gutted him and left him for the animals. He shouldn't have been allowed to live.

"They also taught me about surviving off the land, how to cultivate and grow things."

As soon as he'd been well enough, the first thing he'd done was escape. Without experience or knowledge of the area, he'd wandered off only to end up lost, hungry and alone. When his legs finally gave out, he was grateful, thinking he would soon join his father after all.

But Darrek found him, tracked him down to bring him back. To force him to live despite whatever plans Torren had for himself. He never did understand why the man had bothered.

"They kept me alive and safe." Despite his deep-seated intentions otherwise.

"I'm glad they were there for you. I knew El would look after you if He was able."

Torren felt his shoulders twinge, not agreeing in the least.

"He spared you, and I'm sure He also guided you to find the Vassal so you could finish your father's task." Her voice was full of joy. "And in so doing, He brought you back to me."

Torren dared say nothing.

"Come, let us give our thanks to Him together." Zelene took his cold hand. He didn't resist as she led the way, though he didn't want to do this. She took him into the interior of the house, paintings and pottery he passed flashing recognition in his memories, then to the center-most room—to the El-at.

The circular area, open to the sky, contained a lush and carefully groomed garden. This was the way most of the chapels to El were set up, not like the closed-off room at the embassy. In the center, El's symbol sat on a tall pedestal. Torren had spent many hours here, tending the garden, praying, laughing with his parents. He crushed the memories as thoroughly and violently as he could.

Leaving her sandals at the door, Zelene entered the garden and knelt before the golden icon, pulling him down beside her.

Trying to distract himself as she prayed, he wondered what his foster parents would have made of the temple and the other buildings and art of the island. Darrek had offered more than once to take him to one of the embassies so he could be reunited with his real family, but Torren had refused. Alain was more than happy for it, though she would have supported him if he decided to return.

Once he had begun his travels and joined up with Sal's company, he'd sent gifts and money back to them but visited infrequently. He knew they would have preferred otherwise, yet he couldn't find it in him to go more often. His mixed feelings about their saving his life wouldn't allow it. Still, he'd felt pain when he heard of Darrek's death, and a deep ache as Alain followed not long after.

He'd given up on having a family, yet here he was home again, a mother he'd all but forgotten kneeling beside him.

When Zelene finally raised her head, she glanced over at him and smiled. "I'm so glad you're back." A little color flushed her cheeks. "When Aen awakens again, then everything will be perfect."

He helped her to her feet, saying nothing, knowing once Larana awakened, if ever, he would be on his way once more.

"Lii moved in with me once her husband passed on. She's been a great help." Zelene stared at the tiled floor as she led the way back. "After you and your father were lost, I didn't do too well for a time. But with El's support, I finally decided to push past the loss and to do what I could to promote El's will and your father's good works. So, I followed in Lar's footsteps until I was worthy and could take his place on the council." Her fine-boned hand sought his. "Things have been hard for everyone for some time. The spirit of our people suffered a terrible blow when Aen was taken. It was made worse by the loss of your father and the others. It's been very trying." Her expression brightened as they reached an uncomfortably familiar doorway. "Do you remember this place? It's your old room. I thought you might want to see it."

She pulled aside the curtain over the doorway so he

could step inside.

Diffused sunlight filtered in through small windows and revealed a room frozen in time. Torren felt his breath catch in his throat as the memory washed over him, making him dizzy.

Before him, looking as fresh as when it'd been originally painted, was the mural on the back wall of two stags clashing in battle over a prospective mate. He had begged his father to commission it for him, astounded by the power and brazen stubbornness he'd seen when watching a real pair battling near one of the high mountain outposts of the Chosen. The blue-eyed stare from the larger buck sent a chill down his spine as it followed him about the room—a piece of artistry by the painter that had always amused him, until now.

Tearing his eyes away from the mural, he found his small bed still sat in the middle of the room, his clothes chest at its head. His carved toy figures of Lander monsters were neatly stacked in a corner. His loop, his paints, study scrolls and all the other things that had defined him were just as he'd left them all those years ago. A half-finished vase kept vigil over it all from the top of a small potter's wheel. He'd meant for it to be a gift for the new Aen, if his efforts were deemed worthy, but the child was taken before he'd been able to get it just right.

Torren took a step back, a shudder racking through him. It was as if he'd never left, as if the boy he'd been would be coming back, though the one who'd loved this room and all these things was long, long dead. That boy died when he'd watched his father's blood pour from his mouth, the end of a sword protruding from his abdomen. That boy died when he survived when everyone else lay dead.

"Torren?" Zelene was at his side, an anxious expression on her face.

Bitterness and other less welcome emotions flushed through him. He closed his eyes, trying desperately to center himself. None of this meant anything. It didn't matter. It didn't matter at all.

"I'm all right." Abruptly, he pulled away from her, away from her concern—concern for someone who no longer existed. He half-staggered back out into the hall. How could Mallean have left him here? Why had he allowed this? He needed to get out.

"Torren!"

The panic in his mother's voice so closely mirrored his own it brought him to a stop. She hovered beside him, not quite daring to touch him. He could see the pain in her eyes at the thought she'd inadvertently hurt him. There'd already been so much pain.

He reached for her, his eyes and throat aching, trying hard to pretend his hands were shaking. "I'm sorry. It's just been a long day, and there've been so many surprises."

"No, the fault is mine. I wasn't thinking." Her eyes wouldn't meet his. "All of this has been hardest on you. Tell me what would be best."

He stared at her, sensing she was prepared to let him go if it was what he wanted. And though it was, he just couldn't bring himself to add to her misery.

"Why...Why don't we sit out in the sorium. It...seemed comfortable there. Maybe we could talk."

Zelene gazed into his face, her eyes filled with gratitude. "Yes, yes, let's do that."

They moved to one of the outer areas of the house, where several reclining chairs were set out for relaxation. He took one but didn't lean back, too tense at the moment to even pretend to relax. His mother didn't sit at all.

The silence grew heavy.

Torren shifted in his seat, feeling even more uncomfortable. He fished for something to say, anything to break the tension from before. "I–I noticed the Chosen seemed more armed than I remember."

"Yes!" Zelene sounded grateful for the distraction. "After Aen's disappearance and that of your father's search party, no one knew what might happen next. We all felt very exposed, and many thought it a way to relieve the people's fear." She paced before him. "El had always warned us of potential danger through the Vassal. Now there was no one to give us His word, no way for Him to communicate with us. Some felt taking up arms would prove we were more self-reliant, that this was all a test for us. I disagree, but I can't say it hasn't helped some deal with our loss.

"Still, it sometimes feels as if we've sacrificed something because of it, and in no way has it truly helped Aen in the end." She stopped and turned to look at her son. "Torren, are you...are you here to try and help? Is this why you changed your mind and came back to us?"

He glanced away, unsure how she would take his answer. "Micca thought I might be able to do something. He thought I might be able to find out what happened and help fix it."

"Because you know the Landers," she guessed.

He nodded slowly, not wanting to explain.

"If anyone can help, I'm sure it's you." Her voice picked up animated excitement. "El works through you. He guided you to Aen so you could bring her back to us. I know He will guide you now to help save her."

He sighed, the faith in her words making him ache. There was no divine guidance here.

"How was it the capital was over Caeldanage at this time of year? Wasn't it early?"

Zelene at last sat on the chair next to his. "Yes, it was early. And we stayed a lot longer than usual."

He stared at her, a prickling in the back of his neck. "Do you know why?"

She nodded. "After all our years of prayer, El was finally able to send us a sign by indirect means. The councilors and ambassadors used what contacts they had amongst the Landers and asked them to keep watch for certain signs in exchange for 'favors.' One of them finally stumbled unto something, though just about everyone else had given up hope."

"Oh? Information came from a Lander?"

Zelene nodded. "Seemingly, yes. I don't think anyone specific was named, but we were told they'd seen her and they were making preparations to retrieve her for us."

"And on just this, the island's path was changed and all of you went there?" He was having a hard time believing this.

"Valerian wouldn't have informed the council if he

hadn't been certain the information could be trusted," Zelene stated. "So, you can imagine the council's delight when it ended up that it was you, a Chosen, rather than a Lander who actually brought Aen back to us."

He frowned. So, she'd been expected. But how could anyone in the council have gotten wind she was coming back? Surely, those men in black hadn't been planning to give her back to the Chosen. Had they? It made no sense. Or did it?

He shook his head, not liking the path his thoughts were taking. He decided to change topics yet again. "My uncle's name was Vennel?"

A touch of joy flashed across Zelene's face. "Yes! That's right."

They spent the next while talking about people Torren could barely remember, his mother eagerly trying to help him rekindle what memories she could. She told him of her own life, and Lii's, but didn't ask about his as a Lander. He didn't volunteer the information, and her not asking suited him just fine.

He finally could feel himself starting to relax as night descended outside. Zelene lit a couple of braziers to give them a little light just before Lii appeared with dinner.

Engulfed by aromas half-remembered as she set the dishes down, he listened as his aunt picked up where his mother had left off and filled him in on general doings of the Chosen—who the promising artists were that year, the astounding pottery coming from one of the other islands, which Lander areas seemed to be coveting what types of their work and feathers. She even went into some detail about the plans some were bandying about to properly welcome Aen back during the celebration of El's return to the First Mother in a few moons' time.

He let the chatter wash over him, reveling more than he would have expected in the meal. As Lii had recalled, he did love valmion tarts. She seemed to get no end of pleasure as she watched him reach for another.

"Torren, what are Landers like? Are all the stories true?"

His aunt's innocent questions gave him pause. He hesitated a long moment before answering.

"In most things, I'd say they're really the same as the Chosen. They laugh, they cry, they love. Some actually have quite a sense of humor." He looked down at his plate. "Their lives are harder. Their choices more limited, though a few places prosper as well as the Chosen do. I don't know if it's because their lives are more difficult or because of the things some of them fear, but Landers seem to possess some emotions that are darker than the Chosen's. Still, I believe there are things we could learn from them."

He looked up to find his aunt staring at him wide-eyed, as if not having expected what he'd just said.

"Tell us more."

Much to his own astonishment, he did, encouraged by questions from his aunt and mother. Another couple of hours passed by before he was even aware of it.

"Why, I never." Lii slowly shook her head, as if having trouble assimilating all he had said. "Oh, it's late. I guess we've kept you talking, haven't we? Should I bring some clean linen to your old room?"

Torren stiffened, not liking the idea of returning to those memories again.

"We have a guestroom, if you'd prefer that instead," Zelene quickly suggested.

He slowly shook his head. "My things are at the Vassal's house. Until matters get resolved, it might be best if I stayed there." He saw Lii frown, not understanding his meaning; and he realized his blunder. His aunt didn't know about Larana's true condition.

Zelene pressed on, not giving her a chance to ask what he meant. "If you have time, could we count on your joining us for lunch tomorrow?"

"Of course." He said it with as much enthusiasm as he could muster. "It's a promise."

Both women's faces lit up. He stood to go. Zelene and Lii rose to their feet as well.

"It's been wonderful to have you home again, my son."

Both gave him heartfelt hugs, and he returned them, his heart at odds with itself. "It's...It's been good to see the two of you, as well. Goodnight."

"I'll be sure to make some more tarts for you tomorrow!"

Torren half-smiled at the promise. "I'll be looking forward to it."

With that, he walked through the sheer curtain leading to the exterior of the house. The deepening night's breeze caught him as he moved briskly down the steps, and it felt good against his warm face. He'd survived the evening, despite his misgivings. And though he knew sooner or later questions would be asked he didn't want to answer, he wasn't as unhappy at the situation as he had thought he would be.

All three moons—sisters, in Lander lore—shone above him, softly illuminating the way. Torches burned here and there, lighting the area for the guards up in the spires. Hesitating a moment to get his bearings, he started off in what he hoped was the right direction to the Vassal's home.

As he strolled along the stone paths with their low, trimmed hedges, he replayed all that had happened during the day in his mind. With any luck, Sal would have found out something on those men by the time they reached Caeldanage. He had a feeling they might be the key to the answers they were all looking for.

He didn't notice four shadows sweep over him until a Flyer landed directly in his path. Torren's eyes locked on him, wondering if it was someone else who thought they'd known him. He realized almost immediately this wasn't the case as three others landed around him.

Boxed in, he instinctively reached for his sword. His hand closed on empty air as he remembered he'd left it back in his room at the Vassal's home.

"So, look who we have here."

Torren turned to the side, recognizing Elon's voice. "What do you want?"

The young Flyer gave him an unpleasant smile. "We're doing our duty to our God, as we should. Though I very much doubt it's what you're doing."

Torren glanced upward toward the nearest spire, looking for the guard, but found the spot empty.

Another of the four laughed. "Tyo, I think he's looking for you."

Glancing over his shoulder, Torren saw the one directly behind him was wearing armor and was armed. He cursed to himself. He had the knife in his boot, but it would be hard to reach it in such close quarters.

"I ask you again, what is it you want?"

"I've already told you," Elon said, smiling widely, "we're here to do El's work." The smile died. "You may think you've fooled everyone, but you're wrong. Some of those old fools may be willing to believe anything just to allay their fears, but there are those of us who'd rather reveal the truth instead."

Torren felt himself stiffen. "And what truth might that be?"

"Why, the truth about you, of course. We're here to reveal you for the impostor you are." As Elon spoke, all four Chosen took a step closer to him. "Take that off."

Elon pointed to his vest.

Slowly moving as if to comply, Torren got ready to attack. As soon as he slipped one arm from the vest, he lunged right and elbowed the Flyer there in the ribs, knocking him back. He got a kick in on Elon before a blow from behind sent him to his knees. Before he could make his vision stop spinning and struggle back to his feet, all four Flyers closed in on him and wrestled him to the hard ground.

The stone walkway bit into Torren's cheek as he struggled to free himself.

"Let go of me!"

His pulse thundered at his temples, anger and frustration growing inside him. Laughing, Elon made sure the others had him firmly pinned before reaching into his drape to withdraw a small knife.

"Now we'll reveal you for the liar you are."

As the Flyer bent over him, Torren howled and put every last ounce of strength he could into trying to break free. His captors held tight. Elon only laughed harder.

Pressing Torren's face down with his hand, Elon used the knife to snag his loose-fitting shirt and cut it down the middle of the back. With Torren still uselessly struggling beneath him, he then moved aside both halves of the shirt to reveal what lay beneath.

The haughty, triumphant smile on the young Flyer's face faltered and died. With a gasp, he straightened, stepping back, the knife in his hand falling with a clatter to the ground.

"No, it can't be...He said..."

Torren was suddenly free as the others loosened their holds in their amazement. He shot up from the ground, his fist pulled back. With a grimace of pure hatred, he let fly at Elon's shocked face. His fist connected with the Flyer's jaw and sent him sprawling.

With a sneer, Torren strode up to tower over him. "So, did you get an eyeful? Did you see what you wanted to see?"

He turned scathing eyes on the others as they backed away in panic.

"Please, we honestly thought..." The guard shut up as Torren snarled at him.

Every cell in his body begged him to take out his knife. He wanted to hurt them, hurt them badly. How dare they expose his shame?

"Get out here. Get out of here before I decide to kill you."

Two of them immediately turned and took flight. The third, the one in armor, backed up but didn't leave, staring with concern at the still-stunned Elon.

Torren turned around, ignoring him, telling himself the guard had cause for concern. He swayed, his fists coiled at his sides, staring at the hapless Elon. All he wanted to do was beat the Flyer to within an inch of his life. He was having to resist with all he possessed not to give in to the impulse.

Elon only stared at him, terror turning his face white. "No, no, this isn't right. You were...You were supposed to be...The Black Lords..."

Torren's anger inched back a moment at the unexpected but familiar name. "What did you say?"

Elon's eyes widened, as if he had only just realized he'd spoken out loud. His wings quivering, he scrambled hastily to his feet. "I...I...Nothing. I said nothing!"

Before Torren could stop him, he took off and soared out of reach.

"We're sorry."

Torren turned around just in time to see the guard dis-appear as well. Bastards! Seething, he yanked his vest from where it'd fallen on the ground and slipped it on to hide the scars on his back. Now totally alert and watching everywhere at once, he stomped the rest of the way back to the Vassal's home.

# CHAPTER 19

Micca glanced up from where he sat reading a scroll as Torren came into the room. He set the scroll down and stood, astonishment on his face as he saw his friend's rumpled state and the scratches on his face.

"Torren, what…?"

The rest of the question died on Micca's lips as Torren threw him a furious glance. The Flyer was silent, his expression of surprise turning to concern as Torren angrily yanked his pack from the corner and retrieved a change of clothes.

As he passed on his way to the bathing room next door, Micca was finally able to find his voice again.

"What happened?"

Torren ignored him. When Micca made as if to follow him, he glared until he stopped. The Flyer wisely decided to stay where he was.

Torren bathed with his front facing the doorway, not sure Micca wouldn't barge in. He didn't care if the Flyer saw the few scars he'd picked up over the years in training and combat, but the ones on his back—those were another matter entirely.

When he returned to the room, Micca peeked at him from his bed. Torren ignored him, going to his side of the room. After dumping his clothes, he retrieved his blankets and spread them out on the floor.

"If you're hurt, I can fetch a healer."

"I'm fine." Though the bath had helped take the sting out of the night's events, he was still far from happy. "Nothing happened. Goodnight."

He lay down on the floor with his back to Micca, cutting off further conversation. It took him a long time after his roommate quietly turned off the lamps to go to sleep. Once he finally did drift off into oblivion, he slept without dreams.

# CHAPTER 2

MALLEAN MET THEM FOR BREAKFAST THE NEXT MORN-
ing. Torren made sure to bring his dagger as well as his
boot knife, slipping the former into his belt for easy access. While they ate, Tyleen appeared just long enough to
let them know there'd been no change in Larana's condition during the night. She tentatively smiled at them until she caught sight of Torren's deep frown and then scurried away.

"I'm sure Valerian will call for a vote this morning,"
Mallean informed them as they finished the meal. "It's
very difficult to tell which way it will go."

"How long might the council decide to keep this a secret?" Micca wondered. "Surely, if she doesn't wake up by
the time we reach the Lander city, everyone will realize
there's something wrong when she continues to be unseen."

Mallean slowly shook her head. "Some will want to in-
sist nothing is wrong even when there's no doubt of it.
Only one thing will keep matters from fragmenting into a
situation no one will be able to control."

The councilor sighed, looking suddenly worn.

"We'll find a way to help her," Micca protested. "We
must." He glanced at Torren for support. "I'm sure once
we hear from my uncle and acquire the information
Torren's friend has gathered, we'll be that much closer to
a solution."

The table fell quiet. Before it could become awkward,
Torren asked, "Have either of you ever heard of the Black
Lords?"

"The Black Lords?" Both Mallean and Micca stared at him, obviously not having heard the name before.

"They're a band of soldiers for hire. They have a bad reputation—some say their own agenda. They're ruthless." Torren had heard of them from other mercenaries who'd had to work with them on campaigns, and from some who'd been on the receiving end of their handiwork.

"Are they important?" Micca asked. "Does it have anything to do with—"

Torren stopped him before he could bring up the events of the previous night. "They might be. I'll have Sal look into it as soon as we get to Caeldanage."

Not long after, they left for the council chamber. Zelene beat them there and all but jumped to her feet as she spotted her son coming in.

"Good morning." She smiled up at Torren, taking his hand between hers as he sat beside her. Hers were cold and trembling. Had she thought he wouldn't come? That he'd left again? He tried not to think about this, knowing eventually it would end up being the case.

Micca and Mallean sat on the bench behind them as they'd done before.

While they waited for the rest of the councilors to arrive, Torren studied the faces of the Chosen. Most looked tired, anxious, almost reluctant to be there. They all knew what was on the docket this morning.

Though he searched for it, there was one face he didn't see. Still angry at Elon's attack, he wondered if the Flyer hadn't come because he feared his deeds would be exposed to the council—or that the bruise on his jaw the size of Torren's fist would bring a lot of questions.

Perhaps it had something to do with his slip about the Black Lords. Torren was pretty confident that, however they were involved, the Black Lords weren't taking their orders from the young Flyer. His impetuousness and age ruled him out as an instigator, since all these events spanned nearly two decades. There was no doubt, however, the Flyer was in some way involved with this business.

Elon still hadn't turned up when Valerian ordered the doors to the council chamber closed.

"Greetings, councilors." He stepped out to the middle of the floor, looking stern. "As we do not want a repeat of yesterday, and as we have less than a day before reaching the Lander city, I propose bringing the question of informing the people of the truth about Aen's condition to a vote. Any objections?"

Murmurs sounded in the back, but no objections were raised beneath his scornful gaze.

"Good. When you are ready, you may cast your stones."

He stepped out of the circle as an assistant quickly passed out three baskets, one for each tier of seats. Each basket was filled with small light-gray velvet sacks. Each councilor took one, passing the basket on until it'd made a full circuit. As soon as the empty baskets were gathered, the sacks were opened.

Torren glanced down at his mother's and saw her pull out two flat, smooth stones; one was brown, the other blue. Without pause, Zelene picked the blue one and tossed it into the center of the room.

One by one, all the other councilors followed suit. When they were done, the number of brown stones appeared to be equal to the number of blue ones.

"It is too close to call by color alone," Valerian stated. "A count will now begin."

The room was tense as the stones were separated and counted. Finally, Valerian stepped forward.

"The count is thirty-seven for, forty-two against." His expression was strangely neutral. "The people will not yet be told of Aen's condition."

Relieved sighs and dissatisfied murmurs trickled through the room.

"The Lander city will be reached sometime in the morning tomorrow. If Aen's condition does not change by the following day, we will need to convene again."

No one said anything.

"I would suggest we all spend what time we can praying for success." His gaze scoured the room. "Meeting adjourned."

The councilors and their aides got up to go. A few mingled in quiet discussion, but others seemed eager to get out into fresher air.

Once they made it outside, Torren spotted Valerian exchanging a few words with a couple of the older councilors. As he turned away from them, he moved to intercept the man, knowing he'd find no better opportunity to try and talk to him.

"Excuse me, Tel Valerian."

Valerian turned in the air, already several feet off the ground, his strong wings flapping without effort as he gazed down at him. "Yes?"

"Might I speak with you for a moment?"

Valerian's frost-blue eyes met his impassively, his expression veiled. Slowly, he drifted back to the ground. "What can I do for you, son of Lar?"

Torren tried not to be intimidated by his tone and station. "I had a couple of points of confusion I'd hoped you'd be able to help me with, if you would."

Valerian stared at him long and hard, then looked around at those surreptitiously watching what they were doing.

"Let's take a walk, shall we?"

Not waiting for his response, the councilor set off briskly down the nearest walk. Soon, they were out of the other councilors' immediate view.

For several minutes, Torren kept silent as Valerian continued at a fast yet not hurried pace. They passed an old woman working diligently on some bushes next to the walk. She gave Valerian an acknowledging nod but Torren a gap-toothed smile and a small wave. Feeling a little awkward, he returned it.

Not once in all the years he'd been separated from the Chosen had he thought any of them would accept him again in any way. He was a cripple; he had survived when those better than him had not. Last night's fiasco proved some did feel him to be less than he'd been, but the fact he'd been the one to return Aen appeared to have erased some of the stigma that should have been his at his mode of dress and his lack of wings. How long this friendly attitude would last, especially if Larana didn't awaken, was yet to be seen.

Torren pushed these thoughts aside as they reached the deep shade of a large wisteria tree. Purple blossoms still hung from the long, thin branches, perfuming the air with their delicate scent.

"So, what is it you desired to ask me?" Valerian turned to face him, his imposing figure looming.

Now more than ever, Torren had the distinct feeling this man didn't like him. Somehow, he doubted what he wanted to ask would make the councilor feel any warmer toward him.

"It's my understanding you were the one who convinced the council to detour the capital over to Caeldanage—that you believed the Vassal would be found there. Is this correct?"

Valerian gave him a small, humorless smile. "Yes, you might say that's true."

"How did you know she would be there? It seems strange to me that the information could have come to you, since no one should have known where she was, especially since Aen didn't know who she was herself."

Valerian didn't even blink. "Actually, I'm not the one who discovered where Aen was to be found. The information came to us from a Lander merchant, a contact of Symeas. He'd dealt with him before. As you may not know, Symeas spent some time as an ambassador and had dealings, as they all do, with merchants the Chosen and Lander governments mutually agreed would be allowed to do business with us."

"Yes, but why would you place trust in the word of a Lander?" Torren asked with true curiosity.

"No lead was to be left unchecked where the Vassal was concerned. Our duty demanded it be verified. And nothing better had turned up in some time."

"Still," he couldn't help asking, "wasn't turning the capital around a bit much? There's an embassy in Caeldanage. The ambassador there appears more than capable."

Valerian smile grew wider, yet even less amused than before. "The choice to move the capital was the council's decision, not mine. It was considered to be a show of our commitment to El that we wanted to right what had gone wrong."

Desperate after all these years not to be abandoned by

a god who could no longer talk to them, Torren could see the Chosen's need to cling to any possibility of returning things to what they'd been. But still...

"Might I ask you something?"

Torren was caught off-guard; and though he didn't like it, since he wasn't sure what to expect, he could find no way to refuse.

"Of course." His stomach tightened.

Valerian's frost-blue eyes suddenly bored right through him. "Can you honestly tell me you do not still want to see their blood spilled for what they did to you, son of Lar? For the fact they've made you into one of them? Or are there other reasons for your seeming understanding of the Lander race?"

He tensed. Why was Valerian asking him this? "My feelings one way or another don't change the facts. I don't know who assaulted my father's party and not all Landers were responsible."

Valerian's expression turned blank. "So you say. And yet, I note you still haven't answered my question."

His gaze was even more intense than before.

Torren felt his cheeks grow hot, betraying him. He'd wanted to spill Lander blood, to see it run and stain the ground as his had done. And he had. It was to do this he'd left those who'd helped him, once he was old and well enough, and become a mercenary.

But he'd found no satisfaction in it. It hadn't brought his father back. It hadn't made up for the loss of his wings. And due to Sal's stubborn intervention, it hadn't even brought the death he'd thought might be better to have than to forever live with the memory of what he'd once been.

"No, I don't want to spill their blood. But I will if it's to defend me or mine."

Valerian raised a brow. "I see." Was that disappointment in his eyes? "If you have no further questions, I have other business to attend to."

Before Torren could say anything, Valerian rose on his wings and flew out of sight.

Wondering at the strange conversation, Torren slowly made his way back to where the others were still waiting

for him. Zelene smiled as he joined her and told him Mallean and Micca would be with them for the midday meal.
Lii had promised a feast, and she was sure the two of
them couldn't possibly eat it all.

Not relishing the thought of returning to his old home
alone, he was relieved the others would also be coming.

When they arrived, Lii greeted them warmly. She
glowed with happiness at being able to serve such a large
and illustrious party. She brought food out until none of
them could eat any more.

"Zelene, I didn't know your sister was such a gifted
cook." Mallean gave a contented sigh as she leaned back
in her chair.

"Oh, please!" Lii giggled as if she were a girl, her
cheeks coloring at the compliment.

"Our family has normally shown a flair for one skill or
another," Zelene said, "though in my case, I'm afraid my
sister got all of mine."

"Oh, no, mine is modest by far. Cooking is easy, and I'm
not the one who's risen to the position of councilor," Lii
countered. "Even Torren was showing some promising
skills. He possessed a true flair for art, especially in pottery." She turned to look at her nephew. "Landers do that,
don't they? Did you pursue either one with them? I'm sure
their techniques, though, are probably different from
ours."

He didn't look at his aunt as he made himself answer.
"Yes, they do have a broad spectrum of techniques and
styles. But no, I didn't pursue those things there."

His neutral tone made her eyes grow suddenly sober.
"Oh, I'm sorry to hear it."

"What...What interests occupied you, then, my son?"
Zelene didn't look at him as she asked. It was almost as if
she felt she shouldn't ask, but hadn't been able to help
herself.

Torren told himself he should be used to unwanted
questions by now. "I learned how to till the land, to care
for farm animals, how to make cheese and spin thread. I
learned how to track animals and trap them for food, and
what things are good and bad to eat in the wild. I even
learned about herbs and how to use them."

Darrek had started teaching him the last few after the time he'd tried to run away. As the old farmer put it, he and his wife had gone through too much effort to save him for them to allow him to just run off and die from ignorance. If he were going to leave them, he'd have to have the skills to survive. The memory brought an unexpected half-smile.

"There really wasn't much time for artistic pursuits on the farm, and even less while on campaign."

He realized his error too late.

"On...campaign?" Lii asked.

"Uh, yes, Lander warfare," he answered. "It's how I made a living for a time."

He watched his mother's eyes grow sad.

"You may soon not be the only Chosen to pick such a foreign profession." Mallean said with a knowing look.

Her statement reminded them of the increased martial attitude of some of their people.

"It won't come to that," Lii insisted with conviction. "We have Aen now. Everything will go back to how it used to be."

Neither councilor voiced an opinion.

"How long has Tel Valerian been head of the council?" Torren wanted to know, but also hoped the question would move their thoughts away from subjects better avoided. Larana's hair clip felt cold against his chest.

Mallean answered. "He was chosen by the council two years after Aen was kidnapped, when it'd become obvious her return wasn't assured."

"He's been wonderful," Zelene added. "A pillar of strength for our people. He's kept us on track through all our troubles and setbacks. He's been blessed by El for our benefit."

Micca threw Torren a telling look, knowing as he did what Zelene didn't—that there was a chance Valerian might be involved in Aen's current predicament.

"Aside from you, he is one of the Chosen with the most exposure to Landers. His father often took him along during his time as ambassador, and then later, Valerian volunteered as attache for one of the others before he became old enough to hold the post himself."

Torren nodded, finding the information somewhat at odds with the strange questions Valerian had leveled at him earlier.

"What of Symeas?"

Zelene laughed. "He's had some dealings with Landers as well, and it's been said he's actually spent time with some outside of the usual contractual talks, but this is only due to his great interest in animals and plants. He's always going on about how Landers are wasting the First Mother's gifts." A mischievous smile touched her lips. "He can be quite amusing if you get him talking about those topics. Just don't expect him to stop anytime soon!"

Lunch was delicious and pleasant, Lii making the most of her chance to play host. Even so, it seemed to him there was an undercurrent of tension in the air, affecting everything said or done. It was a measure of the council's fear that rumors of what was truly going on hadn't yet leaked to the general populace. Either that, or the truth was so horrible no one would willingly face it as long as they had a choice.

"I'll have dessert out in a few minutes. Hope you saved some room." With a smile, Lii started collecting the dishes. Offered help, she turned it down. Zelene, however, insisted.

"You've done enough. And since this is my home, too, I should at least share some of the burden."

Not able to dispute this, her sister gave in.

Alone with Micca and Mallean, Torren shifted in his seat to address them.

"Time is running out," he said quietly. "The Vassal won't awaken when we reach Caeldanage, and we're not really any closer to helping her or finding out who's responsible."

Mallean said nothing, but Micca blanched.

"Have we no hope, then? None at all?"

He shook his head. "I didn't say that. But there are some things that need to be set in motion."

"What do you suggest?" Mallean asked.

"Does the council keep documents on where people have been assigned—councilors, ambassadors, their aides?"

She nodded. "Of course. Most are kept here in the Reservoir of Knowledge. As El would have it, its maintenance falls under my jurisdiction. I'm sure I can help you find whatever you require."

Torren glanced in the direction his mother and aunt had gone then looked back at his two companions. "Good. There are a couple of things I want to search for there that may help me get some information I need. Unfortunately, with time being as short as it is, I can't wait until we reach Caeldanage to try and get the rest of it. Micca, if you're willing, or if you know of someone you trust enough to do it, I'm going to need some messages delivered to Sal. The sooner they get to him the better."

"If...If it will help Aen, of course, anything you need." Micca's expression was earnest. "Does it have something to do with the name you asked us about before?"

Before he could answer, Zelene and Lii returned.

"Here it is, as promised." Lii set down a large bowl of fresh fruits and a smaller one full of a thick white concoction. Zelene was right behind her bringing a ladle and several small plates. Lii had a smile on her face as she served each of them the delectable treat.

"Zelene, would you mind very much if I borrowed your son this afternoon?" Mallean asked casually. "I thought I'd get his story on Aen's rescue for the Reservoir of Knowledge."

"Oh, that would be wonderful!" Her eyes lit with pleasure. "I'm sure Aen would wish it. Having a record of it will help give others an understanding in the future of the courage it took, especially when her kidnapping will seem like a far-off fantasy."

Torren looked away, not feeling Larana's return had anything to do with courage.

"You will come join us again for supper, though, won't you, Torren?" Lii waited for his answer with expectation.

"I've put you through so much trouble already."

"No, no, we'd love to have you." She waved his objection away. "Please do join us."

He glanced at his mother and saw his aunt's expectation mirrored there. "Yes, of course. I'll be here."

Their joyous smiles were almost more than he could take. Why did they care so much? He should have been dead to them years ago.

# CHAPTER 21

AFTER DESSERT WAS FINISHED, MICCA, TORREN AND Mallean took their leave and headed off in the direction of the coliseum. Not far from there was one of the few two-story buildings on the island. It was walled rather than open to the outside, and long, narrow windows were cut into the solid stone, glass covering the apertures to protect the contents inside while allowing as much light as possible to enter.

Parchment and paper were the commodities the Chosen didn't produce on their own. Along with metals and some raw goods not found in the high mountains or plateaus, parchment and paper were among the main items the Chosen asked for in trade when contracting to move Lander cargo from place to place.

As Torren entered, the scents of wood, paper and parchment struck his nostrils. Memories flooded him of other times he'd come to this place with his father. He recalled how he'd stared up in awe at shelf after shelf of scrolls containing the knowledge of the Chosen and the wisdom imparted to them by El. He could almost hear his father's soft voice explaining to him the color-coding of the filing system. It sent a shiver of sadness through him.

"You can use the keeper's office," Mallean offered. "If you'll tell us what you're looking for, Micca and I will see what we can find. You'll be able to study the information there at your leisure and in private."

"Thank you, that would work just fine."

Torren waited to make his requests until they'd climbed to the second floor and gone to the back, where there was a small room hidden behind loaded shelves. The room held a small utilitarian desk and chair as well as a convenient shelf holding writing implements, seals and several kinds of paper.

Once inside, he turned to the other two. "What I need, if you can find it, is information on the five councilors who had dinner with Aen that day. I want to find out what posts they held in what Lander cities, or mention of any other time they might have been below for extended periods of time." He noted Mallean didn't appear disturbed, though she was included in the list. "Is it possible?"

Mallean nodded, looking thoughtful. "Yes, I think so. It'll take some work, but between the three of us we should be able to gather the information in a few hours."

"Would it be all right for me to use some of the materials here?" he asked, his mind already moving on to all he needed to ask Sal.

"Yes, please do. In the meantime, Micca and I will go find the bulk of what we'll need."

"Thank you."

Grabbing a sheet of paper, Torren sat down and got to work. He'd finished most of what he wanted to say by the time the others returned, their arms filled with scrolls.

Setting them on the table, Mallean and Micca divided up the scrolls three ways and started going through them, creating a list as they went along. Torren asked her to write down the names they would be looking for, not totally sure he remembered everything he should of the Chosen's written language. He started out slowly; but as the fluid characters grew more and more familiar, his speed increased.

The first time his search accidentally confronted him with his father's name he faltered for a moment. Indulging himself, he read long enough to learn it was an entry listing Lar's service to Waxia, one of the cities in the southern part of the empire. It wasn't something he thought he'd ever known. It felt even stranger to realize *he* had been there not a year or so before. He went back to work.

Every so often, Mallean or Micca would remove some of the scrolls they'd finished with and fetch others. By the time they stopped bringing any more, the light from out-side had grown dim.

Mallean lit a lamp as Micca compiled the three lists. Torren translated the final list into the empire's language. When he was done, he added a few more lines to Sal's let-ter and put the list with it. He then rolled the documents and sealed them with a cord and wax.

"You really think this information will help Aen?" Micca asked as Torren handed the bundle over to him.

He tried his best not to appear too hopeful. "We'll have to wait and see."

Turning away from the Flyer, he gathered up all the papers they'd been taking notes on, including the final list. Using a handy brazier, he burned them to ash.

Micca sighed. "I guess I should get ready to go, then. Uncle Rux will be quite surprised to see me." He gave Torren a half-grin. "I should be able to make it there hours before first light. I'll use the same ruse as before when I go meet your friend."

Torren nodded. "Thanks."

Suddenly, Micca's expression turned uncertain. "Can I...Might I ask a boon of you?"

Surprised by the request, he stared at the young Flyer. "What is it?"

"Will you stay by her side tonight? I'll speak to the guards, and they'll allow it. I just don't feel comfortable leaving her, but if I know you're going to be there with her, it would greatly ease my mind."

Torren felt Larana's hair clip pressing against his chest. "All right."

Micca grasped his arm with heartfelt gratitude. "Thank you."

Torren started toward his mother's place and, to his amazement, found Mallean accompanying him.

"Micca mentioned something to me that was rather cu-rious."

He said nothing, waiting to see where she would go. He caught her looking at him inquisitively from the corner of her eye.

"It seems you came in late last night, looking bedrag-gled and with your face scratched," she went on. "Micca told me you wouldn't talk of it."

"That's true." He felt a taste of his previous anger rise inside him. "And I still won't." He almost glanced down at his belt, where he'd strapped on his dagger that morning. He didn't intend to be caught unprepared again. He wondered if she'd noticed it.

"I see." She continued beside him for several minutes without saying anything. "Was there someone here who decided you were a Lander?"

Torren's gaze snapped over to hers. He cursed, realizing by his very reaction he'd given himself away.

Mallean sighed, looking away. "It is but more evidence of how low we've come. Anyone who bothered to look at you would be able to tell you're one of us. That those of us who knew your father have confirmed your identity should have also been enough. I was astounded when it was questioned at the council meeting.

"Though, now I look back on it, I suppose I shouldn't have been. Our people see Landers hiding in every shadow these days. It bodes ill for our future if Aen does not awaken soon." She glanced up at him, her eyes troubled and sad. "You were greatly wronged, yet kept silent at the meeting, saving us from embarrassment. For this, I thank you. Still, it grieves me we would treat one of our own this way."

"Actually, I've been treated better than I would have ever expected," he admitted. He said nothing else as they arrived at his mother's house.

"You are too kind," Mallean said. "In some ways, you are very much like your father."

He nodded, not wanting to argue the point.

"Enjoy the time with your family. I will see you on the morrow."

"Goodnight." He watched her out of sight then made his way up to the columned outer area of the house.

Conversation at dinner was light, carried mainly by Lii. Torren ate in silence, his thoughts on what might occur the next day. He was sure his mother's thoughts lay there as well, as he noticed she only picked at her food.

As soon as it felt polite to do so, he extricated himself and bade them goodnight. This time, as he made his way back to his room, he didn't allow himself to get lost in his thoughts, instead keeping his senses primed on his surroundings. He needn't have worried; he arrived at the Vassal's home without incident.

With Micca no longer there, their shared room was quiet. After a more leisurely bath than he'd enjoyed the last few days, he took the time to try and repair his torn shirt. Finally, having nothing else to keep him occupied, he made his way toward the main bedroom. Stopping before the covered doorway, he slowly drew the drapery back but didn't go in.

"Excuse me."

One of the two Flyers he had met here before appeared without getting too close. When he saw who it was, the guard moved out of the way to let him in.

"We've been expecting you."

As soon as he entered the dim room, he spotted the second guard hidden in the shadows.

"Tyleen is with her now."

The two made no move to follow as he headed for the curtained side of the room. Pausing there for a moment, he pushed the drapery aside and entered the lit space beyond.

Larana lay in the wide bed, in the same position he'd seen her before. Torren wasn't sure, but he thought she seemed paler than when he'd been here last. How long would it be before she started to waste away, before the ravages of unconsciousness showed visibly to where no one could deny the truth?

"Has there been any change?" He aimed the question to his right where Tyleen stood meekly, trying hard not to call attention to herself.

"No change. But perhaps tomorrow..." Her gloomy expression showed she no more believed it than he did. He said nothing.

"Are you really planning to sleep here?" she asked a moment later, speaking even more softly than before.

"Yes," he answered, still watching Larana. "Will it be a problem?"

The Flyer shook her head, still keeping to the shadows.

"I just wanted to know." She turned to a table behind

her then faced him once more. "I brought some blankets for you."

Torren tried hard to give her a smile, knowing she felt uncomfortable at his intrusion.

"Thank you." He took the blankets and laid them out on the floor beside Larana's bed.

"I–I will leave you, then. If you need anything, tell Mar or Styn and they will send for me." She didn't look him in face.

"I'll remember that."

Half-bowing in his direction, Tyleen made her escape.

He slumped down onto the floor, staring after her, feeling slightly amused. Micca must carry a lot of weight with her; otherwise, he felt sure leaving him alone with the Vassal would have been the last thing Tyleen would be willing to do.

Glancing back toward the bed, he stared at what he could see of Larana's face for a long time. He took her hand, but felt no more of her presence in the touch than he had before. Squeezing it hard for a moment, he set it gently on the bed and turned his back on her, readying himself for sleep.

# CHAPTER 22

Torren opened his eyes, shocked by the stench assaulting him. It was the pungent, nauseating odor of rotting flesh—the smell of death. What he saw did not make him feel any better.

Though he was no stranger to battle or the carnage it left behind, the sheer magnitude of what he confronted gave him pause. Dead horses, dismembered men, discarded weapons covered the ground in a blanket of the slain. As far as he could see was littered, trampled earth stained the deep brown of drying blood.

Looking at the corpses nearest him, he identified the livery of the emperor as well as that of the nations of the Northern Tribes. Hatred twisted their dead features. It rose in a miasma of loathing, buoyed by the shroud of death. It clung to him, seeking him, trying to lodge within him. He felt himself sway in its grip, as if it were a living thing.

Darkness flushed into the valley, making Torren spin around, a gasp escaping his lips, numbing cold shooting through him. Something hovered above him, but he couldn't make out what it was. Triumphant, spiteful laughter trickled down to him.

The sound of wings filled the air before everything suddenly went black.

※

Torren woke staring into the semi-darkness. He sat up, still gripped by the strange dream. With a shiver, he turned to check Larana and found her just the same.

Only his recurring dream had ever disturbed him in this way—had ever seemed so real. That one had followed him until he'd met Larana. Only after he'd left her behind had it ever deviated, and then only to put her in his place.

Now, since arriving here, he'd had two others as vivid. The portent in this last dream seemed clear—the destruction of thousands, tens of thousands of Landers. But why? What would there be to gain from it? Or was he misunderstanding what he had seen?

No longer the least bit sleepy, he wrapped one of the blankets about him and leaned against the wall. He stared at what he could see of Larana, questions spinning in his mind. By the time Tyleen returned at sunrise, he'd gotten no closer to finding any answers.

The morning crawled by as Torren met Mallean for breakfast, most of the dishes remaining untouched. He didn't tell her of his dreams, but perhaps she sensed his mood, for little passed between them.

More and more Flyers walked or flew by the Vassal's home, some looking at them with eager anticipation. Whether they knew the truth or not, a few must have connected the capital's return to Caeldanage to the Vassal. He knew their curiosity would only increase when she remained in seclusion.

Close to midday, the councilors, one by one, arrived at the Vassal's home and gathered behind a curtained-off area in the sorium. Torren noted eyes red from lack of sleep, taut faces filled with tension. A few had hopeful visages, but even these were frayed at the edges. It was what all of them wanted most, but none, it seemed, truly believed Aen would awaken once they reached Caeldanage.

As they milled about in subdued silence, a guard in gold armor rushed in and, spotting Valerian, made his way over to him.

"Sir, we've reached the Lander city."

Agitated wings rustled throughout the room. Valerian acknowledged the information and sent him on his way. He then turned to the others.

"The healer Ryn, myself, Tel Mallean, Tel Icos, Tel

Mides, and Tel Symean will go see her now. We'll report back to you momentarily."

Torren, who'd been hanging out at the periphery of the room all this time, stiffened as he listened to the list. Aside from the healer, the names were the same as those who'd been with Larana when she'd first collapsed. Ducking under the curtain, he ran into the interior of the house.

Following the hallway, he stopped before the drapery and carefully lifted it aside to announce his presence. The two guards inside were not the twins, but they'd obviously been told about him because they gave him no trouble and let him through.

Not lingering, he crossed the room past the hangings dividing it. Barely sparing a glance for Larana, he continued on until he found a deeply shadowed corner to stand in. Only then did he turn to look at her in the gloom. As expected, she was exactly as he'd left her earlier that morning.

He heard voices shortly, whispering on the other side of the curtains. He pressed as close to the wall as he could as the hanging was drawn aside to admit the five councilors and the healer. Quietly, the councilors flanked Ryn as he came to inspect the patient. Torren watched his every move as the healer bent over Larana.

Ryn finished his observations and turned to face the anxiously waiting councilors, his wings and shoulders drooping. "I'm sorry. I can find no change in her condition at all."

All six stood in silence for several breaths. Valerian was the first to find the voice to speak.

"Thank you, Healer."

Ryn bowed slightly and left.

"I suppose this is to be expected," Valerian said once he'd gone.

"What do we do now?" Symeas sounded on the verge of despair.

A resounding thump rang in the room. "We'll do what we have always done, of course!" Icos glared at the rest of them. "We'll do what we can and trust El to guide us. There's nothing else we can do."

Mides nodded. "We'll have to let the people know."

"No," Mallean said, a little forcefully, "let's give it a little more time. It could just be that her recovery will take a while."

"Yes! Maybe she's right." Symeas's face brightened.

Valerian, on the contrary, frowned. "This isn't like you, Mallean."

She avoided meeting his scrutiny. "Perhaps, but it is how I feel. The disaster we will have to deal with once all are told will be bad enough. One more day of waiting should make little difference." She glanced at the others, her face confident. "Besides, we could use the time to pray and also compose how we will break the news to the people."

"There is sense in what she says," Icos concurred. "Some amongst our colleagues will not take well to our news and will need time to come to grips with the truth before they are able to pass the tidings onto others in a calm manner."

"The main thing we should do is to get every last one of us, whether on the islands, in an embassy or on a mountain, to pray. If we speak with one voice once this disaster is revealed, perhaps El would see fit to guide us again." Mides's eyes were full of grief, as if he were trying hard to believe his own words and failing miserably.

Valerian's stare took in each of them in turn, his expression veiled. "That's assuming we're able to control the people at all."

With this pronouncement, they left, their expressions unsure and worried. Torren waited several minutes more then followed.

He considered waiting out the coming storm once the other councilors were informed of the lack of change in Larana's condition by going to his room, but then thought of his mother, out there with the others. He slipped back to the sorium and returned to where the councilors were gathered.

Soft murmurs slipped through the curtains, as well as stifled weeping. He heard Valerian's composed voice speaking, but couldn't make out the words. Not long after, the councilors started leaving in ones and twos. Those with tear-streaked faces walked slowly around the sorium trying to first regain their composure before stepping outside.

Once or twice, Torren caught confused or half-angry looks flashing in his direction. As he wondered what they were about, one of the councilors passing him stumbled. Without thinking, he reached for the old woman's arm to steady her. She looked up with a grateful smile until she spotted his clothes and face.

With a hiss, she pulled away, glaring at him. "Unclean!"

He stared at her, taken aback, as with an expression of revulsion the woman turned and hurried from him as fast as possible, slapping softly at the place where he'd grabbed her arm.

With a flash, he recalled the voice he had heard in the council room speaking of the possibility Larana was tainted and thought this might be the woman who had spoken. How long would it be before more of them started looking at him in the same way? How long would it be before they decided to blame him for Larana's current state?

He was still pondering these dark thoughts when he finally spotted Mallean and Zelene. His mother's eyes were red, her wings half-wrapped about herself.

"Are you all right?" he asked her, concerned.

"Yes." She wiped at her face. "I'm sorry. It was foolish of me, but I'd truly hoped Aen would awaken when we got here."

Mallean gave her a sympathetic smile. "You weren't the only one."

"What has the council decided to do now?"

"We're going to wait, to give El and whoever He might be guiding time." Mallean gave him a knowing look. "Sometime tomorrow, though, depending on what happens, another vote will be called for."

Zelene clasped her hands before her, eyes filled with sadness. "I'll pray with all my heart it won't come to that."

"Mother." He waited until she looked up. It felt strange saying it, almost like a word in a foreign language. "I'll be unable to join you today." He studied her intently, half-afraid the light would go out of her, as it would when he truly left her.

"Oh?" Her brows pinched inward, and her face paled a little, as if she were expecting the worst.

"He's going to be helping me with a line of inquiry a number of us are pursuing to help the Vassal," Mallean confided.

Torren was amazed by her candor, but then understood it would be the least hurtful way to deal with the situation.

"I need him to go down with the messenger who'll explain what we're doing here to Dom Rux."

A look of panic crossed Zelene's face but was quickly hidden.

"Will you be gone long?" She tried hard to make the question sound casual.

"No more than a day, depending on what's happening. I'll try to send word if it'll take longer than that."

"All right." She tried to smile bravely, but it crumbled. She reached for her son and hugged him, wrapping him with her arms and wings. "I love you."

Torren's heart gave a sudden lurch, knowing it would be much worse than this when he told her his final goodbye. His eyes burned.

After several long moments, Zelene let him go. She nodded to both of them, eyes lowered, and left.

He watched her go, his emotions in turmoil. "Thank you."

"Her heart is pure, and she's honorable, like her husband. I felt we could trust her with this."

Once his mother was out of sight, he turned to face Mallean, surprising her in an intense scrutiny focused on him. "Will you be coming as well?"

She shook her head slowly. "No. Though Rux is a close friend, it would seem strange for me to go, especially at this time. Though I am sure whoever is responsible for this is aware we might be doing something, I don't wish to show our hand any more than necessary."

Torren nodded, concurring with her caution. "When can I go?"

"Would now be too soon?" she asked.

"Let me get my weapons, and I'll be ready."

"I'll wait for you here."

He took his time returning to his room, not wanting those still milling about to notice him. Once there, he took off his dagger, replacing it within his pack, and used one of his own blankets to wrap his sword. He eyed his back-pack, tempted to take it along, but then decided against it, knowing it would give too final a look to his leaving. He would just have to do without.

Tucking the wrapped sword under his arm, he returned to find Mallean leaning against one of the outer pillars, watching those flying by.

"I'm ready."

She nodded, turning to face him then leading the way down the steps. "Micca arranged for this before he left. The two who will take you down should already be wait-ing for us."

She strolled along a path as if she were going no place in particular. He bristled inside with impatience but forced himself not to show it, understanding what she was doing. A couple of times Flyers descended to have a word or two with her. Every time this happened, Torren had to keep from flinching, half-expecting an attack.

Eventually, however, they reached the southern tip of the island. A large park took up the area there, a tall ga-zebo close to the edge affording an unimpeded view of the sky and the land below. It was strange seeing an edge to the land. Only a Flyer could feel comfortable looking at the drop there. Torren was glad they were not going too near it as Mallean led him straight toward the gazebo. He could make out two figures already waiting inside.

As they stepped within and the two stood up, Torren immediately recognized them—Mar and Styn, Larana's night bodyguards.

"Did you bring it?" Mallean asked them after a brief greeting.

"Yes, we hid it over there early this morning." One of the brothers—it was hard to tell which was which—pointed toward a hedge not far off.

Torren threw Mallean a questioning glance, not know-ing what she was talking about. She gave him a small smile.

"The twins will be your transport. And to make it easier

on them, I had them borrow Aen's Wings."

"Her wings?" Even as he asked, something about the term seemed familiar to him.

"They'll show you, don't worry. You'll arrive there safely." She reached within one of her sleeves. "Please give this to Dom Rux for me. You can send one of the twins or Micca to see me if you're in need of anything at all."

Torren took the wrapped scroll and tucked it into the blanket with his sword. "I'll make sure he gets it."

"Mar, Styn, I'm putting him and all our hopes in your care."

Both men gave her a half-bow.

"This way," one told Torren.

Mallean remained within the gazebo as he followed the twins out. They went to the hedge, and behind it, he spotted something covered by a large cloth. As they carefully folded the cover, he recalled the meaning of "Aen's Wings."

Before him lay a silver-framed rounded box with large rings on the top. Part of the top was a hatch, which would open to admit the Vassal into a padded and cushioned interior. Once he or she was inside, the hatch was closed and two or more Flyers wearing leather harnesses with strong ropes attached hooked themselves to the large rings on the top and became, in essence, the Vassal's wings. In this way, they would be able to transport him or her to any destination.

"Please get in." One of the twins held out an arm to help him clamber up onto the box. Slipping inside, he tried to get situated in the enclosed space. Small grills cut into the circumference allowed light in and gave a restricted view of the outside. He shifted, trying to get comfortable, the soft silk feeling strange to his callused hands.

A light rap on the hatch a few moments later informed him the twins had donned their harnesses and were ready to go. Feeling more nervous at the strange arrangement than he would have thought, he waited for the box to lift. To his amazement, he barely felt it as they rose off the ground. The box flew forward toward the tip of the island. It wasn't long before they traversed its shimmering protective field.

Watching the view, Torren saw the land drop away. Like a leaf falling off a tree, they drifted toward the bustling city below. He was able to get a glimpse of the city wall as they descended and saw it was packed with guards. Wary eyes turned in their direction while others looked up to see if any more Chosen would come.

As they neared the embassy, he spotted a small crowd by its front gates. He frowned as he noticed men wearing not only the city's livery but, more disturbingly, the emperor's as well, and at least one man there was in state dress. They all stared up as one of the guards pointed them out to the rest.

Not long after, Aen's Wings gently touched down on the embassy's roof. One of the twins opened the hatch, allowing him to stand up and stretch.

Once he got out and the other two discarded their harnesses, all three headed for a trapdoor on the northern side of the roof. Torren took the stairs beneath it two at a time. After a couple of false starts, he and the twins were finally able to find their way to the ambassador's reception room. They'd seen no one on their way down, which he thought unusual, but for the moment didn't question.

The reception room was empty, but Rux's office door stood ajar. Torren headed toward it. When he reached the open doorway, he found the ambassador pacing, a frown creasing his face, his wings drooping.

"Dom Rux?"

Rux looked up, the worry lines on his face momentarily disappearing.

"Torren!" With long strides, the ambassador crossed the room and grasped his arm in greeting. "It's good to see you again." He glanced over his shoulder at Torren's companions. "And the twins. Micca said you two would be the ones bringing him."

"Dom Rux." Both brothers gave him a half-bow.

"Is Micca here?" Torren asked, his impatience getting the better of him.

Rux's worry lines returned. "No, not yet. Though I expect him back at any moment."

"Has something else happened?"

Rux waved the question aside. "Not exactly. Some other problems have arisen."

Torren nodded. "I presume you're referring to the Landers at the gate?"

"Yes." Rux looked suddenly weary. "The governor has come, and with him an emissary from the emperor." He resumed pacing. "When the capital appeared before, I was able to placate the governor's fears. But now that an emissary from the emperor is here and the capital has, as far as they're concerned, so unexpectedly returned…"

Torren finished what he left unsaid. "They've come looking for answers."

"Precisely. And they're growing more impatient by the moment." Rux stopped by the side of his desk and apprehensively picked at a loose feather on one of his wings. "I sent a message to Tel Valerian the minute they appeared asking how the council wanted me to proceed, but I've heard nothing." His troubled gaze met Torren's squarely. "I don't think I'll be able to hold them off much longer without them jumping to some sort of negative conclusion and possibly resorting to violence."

Torren frowned, wondering if things were deteriorating here as rapidly as in the capital. "The council probably won't be much help. Aen didn't awaken as many expected."

The twins nodded in agreement. "He's right, Dom."

Rux sighed deeply.

"I was afraid this might be the case." He came to stand before Torren. "You've worked closely with them. You understand better than any of us what motivates them. I would greatly appreciate it if you would be in attendance when I meet with them."

Torren hesitated, wanting more to get together with Micca and Sal and pool their information than coddle some frightened politician. "I don't know what help I can be to you, but if it's what you want."

"Excellent!" Rux's wings bobbed a little higher. "For, you see, I think more than just our unexpected presence here is on their minds." He glanced over at the twins. "Refreshments for the guests are being prepared in the kitchen. If you would tell the staff they'll be here presently, I'd appreciate it. Please feel free to take anything you want and to use one of the rooms to rest for a while, if you need it. I know you've not slept since keeping vigil over the Vassal last night."

"Thank you, Dom." Grinning, the two usually sober-faced brothers left.

"Let me go advise the guard to let our guests in. I'll only be a moment."

Taking his time, Torren wandered around the office and studied its maps then stepped out into the reception room. He hadn't been there long before Rux returned.

"They'll be here momentarily." He threw Torren an odd look. "The governor is a strange one, even for a Lander. I've not met the emissary before."

He nodded, grateful for the information, then helped Rux move some of the chairs into a loose, nonthreatening circle. They'd finished by the time a knock rang on the reception room's doors.

Rux, wings twitching, strode across to open them himself. Torren watched from the back of the room as a tall, thin man dressed from head to foot in ruffles and lace entered behind a shorter, portlier man in dark velvet. Beyond them, four men, two of the city guard and two in the emperor's livery, stared with almost open distrust at the four armed Flyers keeping them company.

"Gentlemen, please come in and make yourselves comfortable." Rux swept a hand toward the chairs. "I apologize for the delay in seeing you, but it couldn't be avoided."

"Well, it was highly irregular of you, Ambassador. And the heat does so bother my complexion." The taller of the two men sat down, dabbing at his face with a silk handkerchief. The other continued to stand. After a moment, he made a slight noise with his throat and stared hard at the other.

"Oh! Yes, yes, where are my manners?" The governor stood once more. "Ambassador, I want to introduce you to the emperor's emissary, the distinguished Count von Duren. Count, this is Rux, Ambassador of the Chosen for the city of Caeldanage."

The emissary gave Rux a curt bow.

"It is an honor, sir." Von Duren's dark eyes met Rux's then slid from him to Torren. "And might you be the Vassal?"

He laughed before he could think better of it. Von Duren's eyes narrowed dangerously.

"No, I'm sorry, this is not the Vassal." Rux moved to stand closer to Torren, color rising on his neck, the fact he might be mistaken for Aen not having occurred to the Flyer. "This is Torren, the son of one of our councilors. He's been lost to us since he was a boy and has but recently returned."

"Really?" The governor, who'd sat down once more, now moved up to the edge of his seat with unconcealed interest. "It would explain the clothes. But shouldn't he have...?"

Von Duren sent the governor a scathing look even as the latter realized the impropriety of what he'd asked.

"Oh, my. I do apologize." His interest didn't seem to diminish in the least.

Just then there was a soft knock at the door. Before anyone could move to answer it, they swung inwards, revealing the twins, each carrying a loaded tray. The guards, still in the hallway, glanced in as the two strode forward then closed the doors.

Rux smiled in relief at the sight of them. "Gentlemen, how about some refreshments? Please, help yourselves."

The twins lowered the trays before each of the two men and, after they'd made a selection, set them on small side tables within easy reach. The twins nodded in Rux's and Torren's direction and let themselves out.

"Oh, eva fruit! I love how your cook prepares it." The governor glanced over at von Duren. "Do try some. It's fabulous."

As the governor helped himself to more of the fruit, Torren sat down, eyeing their two guests circumspectly. This was the first time he'd met the governor, but it looked as if some of the talk he'd heard about the man had substance. Still, if this was all there was to him, he was sure there'd be no way he would have lasted in his position as long as he had.

Von Duren was another matter—there was no subter-

fuge in him whatsoever. It was hard to tell which one to
watch out for most.

Rux took a piece of fruit and some white cheese and sat
down. He ate his tidbits slowly, waiting for one of his
guests to start the conversation. He didn't have to wait
long.

Von Duren quickly munched down his food, eating just
enough to seem polite before sitting forward and pinning
Rux with his gaze.

"So, Ambassador, are you the type of man who minces
his words or one who enjoys getting down to business?"

The governor looked pained at the bluntness of the
question and dabbed at his lips with his handkerchief.

Rux met the count stare for stare, the beginnings of an
amused grin tugging at his lips.

"Personally, I prefer to speak simply and to the point.
Though niceties, at times, do have their place."

"I see." Von Duren cast a glance in Torren's direction, as
if trying to ascertain the same thing about him. After a
moment, his gaze came to rest once more on Rux. "All
right then, let me be blunt. Why have your people returned here so suddenly? Why was the usual visiting time
accelerated in the first place?"

Rux blinked slowly but didn't look away. "Those are excellent questions, but I don't have the authority to reveal
the answers to you. I can tell you it is an internal matter
and has nothing to do with this city or the empire."

"Hah." The two men stared at one another long and
hard. "Internal the problem may be, but it seems strange
it would be centered not only here but so close to the border."

Rux raised a brow. "What would the border have to do
with anything?"

Torren spoke up for the first time. "There are rumors of
a military buildup in Galt, on the other side of the mountains. It would seem the emperor suspects the Chosen's
being present here at this time might be related."

Von Duren sent him an appreciative grin. "Precisely."

"Keer, this insistence on bluntness is so gauche." The
governor sent Von Duren a hurt look. "It just takes all the
fun out of things."

The emissary ignored him. "So, can you explain this?"

Rux looked away for the first time. "No, I can't. I'd heard some of those rumors from the merchants we deal with but didn't pay much attention to them. Disputes amongst Landers don't normally affect us; and since there's been no evidence of anything untoward, the council hasn't been informed of any of it. It's just a coincidence."

"Maybe not."

Rux sent Torren a startled look. He now held everyone's attention.

"This is purely theory. I have no concrete proof. But there's a chance these matters could be related, though not in the way you might think.

"Some of the internal problems the Chosen have experienced, I believe, have been partly orchestrated by a group of mercenaries called the Black Lords. The very fact they've been active in this area might also have something to do with the rumors of the build-up of arms in the mountains."

Now that he'd voiced his thoughts, he felt more than ever he was on the right track.

"The Black Lords, here?" Von Duren's brow darkened. "You didn't happen to mention this minor tidbit to me, Governor."

A thin, plucked brow rose on the governor's face, stimulated by the displeased tone in the emissary's voice. "I hadn't heard they were hereabouts myself. It's not as if they're causing any obvious trouble to bring attention to themselves. Besides, there've been other things to keep me occupied."

He sent a sly glance in Rux's direction.

Von Duren grunted and shifted his attention back to Torren. "And what, precisely, do you theorize they're doing here?"

He hesitated for a moment, again wishing he'd already gotten the information he'd requested from Micca and Sal. "I would guess they're here for a couple of reasons. The first would be to carry out what they have in mind with regards to the Chosen. The second would be to spread rumors on both sides of the border and at some point instigate incidents to bring the empire and Galt to war."

Dread-filled silence met this statement. Rux stared at Torren, turmoil darkening his face as he reconciled what he knew with what he'd just heard and came to the same conclusions as his dead friend's son.

"What would they hope to gain from this?" The governor looked even paler than usual.

Torren shook his head. "I don't know. Maybe the weakening of the two countries, hopefully leaving them open to be taken over by a third party."

The explanation didn't sound quite right to him even as he said it.

"No one would dare." Von Duren glared at him. "It also wouldn't explain why they're working against the Chosen, the details of which the ambassador has already made clear his reticence to clarify."

"Now, Keer, it's not as if we tell them our secrets."

Von Duren threw the governor a dirty look. The other continued nonplussed.

"Regardless of what we can or can't tell each other, I would suggest looking into the Black Lords. It could be that if we can find out what their connection is with all this, it will answer a lot of our questions."

"That goes without saying." Von Duren's harsh demeanor eased a bit. He turned to Rux. "But don't expect our search for them means you are totally absolved of suspicion. With no more information than you've given, we can't dismiss the possibility there might be more afoot here than we know."

Rux rose slowly to his feet. "My people have never given yours cause to doubt us in any way before. Why would you doubt us now?"

Von Duren smiled at the question, leaning back in his chair. "Things haven't been as they were with your people for a hand-span or two of years. The empire is aware weapons have been purchased by your race in excess of previous periods. The term 'grub' has come more into common usage.

"Secrets are being kept from us as well. There has been a general change of attitude that hasn't been favorable to us at all, not that it has ever been all that good."

Rux glanced away, color once more rising on his neck. "You're right, there has been a change, but if our current problem is resolved, we should be able to go back to where we once were. The Black Lords could well be involved in the very matters that have brought us to this current state."

"Are you saying these men might be holding the Vassal?" Von Duren's eyes gleamed.

"N—No." Rux's surprise was clear. He shot a look in Torren's direction, one von Duren didn't miss. "I'm sorry, but I have no leave to speak to you of these matters, as I said before. Please try to understand my position."

Von Duren slightly inclined his head. "Of course." He glanced slyly in the governor's direction. "I guess we're done here then." Both men stood. "Please don't hesitate to call on us if some information comes to light that you *can* share."

"I'll do that, of course," Rux said. "And at the same time, I would hope you will keep us informed if you happen upon anything."

"You can count on it," the governor agreed cheerily. Von Duren said nothing.

Torren followed Rux as he accompanied the two men to the door. He watched them as they gathered up their guards and took their leave.

Once they were gone, Rux sighed with relief.

"I'm glad that's over." His eyes were troubled, but after a moment they cleared and he gave Torren a half-smile. "The emissary appears sharper than most men sent to deal with us. They must be truly worried about matters as they stand."

Torren nodded. "They're afraid if a conflict does break out, the Chosen will pick one side or the other, or that you've already picked one. If the Chosen weren't to stay neutral, as they've done in the past, it could mean assured supplies for one side and not the other, and this could very well decide the war. And if you actually joined in it..."

Rux's wings shuddered. "It frightens me to think such a thing could happen. Ten years ago it would have been unthinkable. We would have never contemplated meddling in Lander affairs. But with things as they stand now, half our people thinking Aen's disappearance was their doing, and if the Vassal doesn't recover...."

"Are they gone?" A blond head peeked out from Rux's office, startling them.

"Micca! When did you get back?"

Sheepishly, the young Flyer stepped out of the office. "Not too long ago. I would have been here earlier if not for the mob at the gates. When we finally could get in, we just used the back stairs and came down from there."

"We?"

Micca grinned. "Yes, I brought a friend of yours along with me."

Sal stepped out, looking more subdued than was his usual wont. "Hello."

He bowed in their direction. Torren wasn't sure if he should be astounded or amused.

Looking more closely at the two, he noticed dark circles under their eyes and their sagging, tired stances. It appeared they'd been quite busy through the night and morning.

"Who were those men, Uncle? One of their voices sounded familiar." Micca headed for the nearest chair and flopped down on it, his wings lying half on the floor. Sal took the one next to him, sighing in relief.

"It was the governor and an emissary from the emperor," Rux replied. "They wished to know why the island has returned."

Sal nodded. "I'd heard rumors someone was sent. Brought a whole lot of troops with him, too. Got here not long after you left, Torren. Everyone's gotten a lot more on edge about things since then."

"So, you weren't able to find out anything?" Torren felt his chest grow tight.

"On the contrary." Sal broke into a grin. "I'd say things were actually quite fruitful. Right, Micca?"

The Flyer's face also looked bright.

Torren waited with a quickly rising brow as neither said anything. "And you're going to share this bounty with me when?"

Sal's grin changed to a full-fledged smile as he removed

a number of papers from his jacket. Some of them were from the letter Torren had sent.

"The information Micca brought from you was quite helpful," he stated. "Some of my contacts were able to give me the name of a guy who's into some of the shadier businesses in town, but who'd be willing to provide information for a price. Everyone else I'd asked about it didn't know what to make of it, but his line of work had actually brought him across it before."

"He identified the poison!" Micca blurted out, his excited gaze swapping between Torren and Rux. "It comes from the mountains, from a spiny fish that grows in some remote lakes up there."

"The guy seemed to think whoever gave it to her was damn lucky not to have killed her," Sal added. "The fish are rare, but a little goes a long way. It's also expensive and, as we found out, not well known."

"Is there some way to counter it?"

Micca's face lost some of its light, and for a moment his very real exhaustion came to the fore—but only for a moment. "He wasn't sure. He said he'd never needed one. But he did give us the name and location of someone who would know."

"Here in Caeldanage?" Rux asked before Torren could.

Sal shook his head. "No, across the border, off the high point of the pass. There's an ex-priestess of the First Mother there by the name of Mala. Seems she deals in rare herbs and other things. I got the feeling our guy might be a little afraid of her, which you never see in someone in his line of business.

"He said she was once with the temple here in the city, and had even been invited to the capital, but refused. Soon after, she headed off to the mountains. Priests of the First Mother are a weird lot, with their beliefs in neutrality, not pushing for things one way or the other, just letting them happen. Can't say for sure they're all that way..." He gave them a tired grin. "...but this one might be the real thing.

"Anyway, our contact was sure she'd know if there was a cure, especially since she's just about the only source for the poison in the first place."

"What else did you find out?" Torren sat down, his thoughts racing.

Sal shuffled the papers in his hands. "Well, I was able to get enough information to make a rough outline of where the Black Lords have been off and on in the last twenty years. It's very sketchy, and not all too reliable, but I was able to get enough to make some matches with the dates you sent me."

He handed Torren back his original letter as well as several other rough sheets of paper.

"So, when are we leaving?"

Torren looked up distractedly from scanning the notes. "Leaving?"

Micca leapt to his feet, almost falling in his haste. "Yes! We have to go get the antidote for Aen. Everything else should take second place to that. We should leave immediately."

Torren shook his head. "It won't be immediately, though I agree with you we should leave as soon as possible." Micca started to protest, but he cut him off. "If we don't do this right, the wrong people will find out what we're up to and stop us. Is that what you want?"

Deflated, the Flyer shook his head and sat back down.

"What will you need? Ask and it shall be provided." Rux's eyes gleamed.

Torren gave him a grateful nod. "Sal, Micca, let me and Dom Rux take care of things on this end for now. You two are barely able to move, so a couple of hours of sleep would seem in order. I promise you we'll set out as soon as possible."

"It *was* a long night," Sal admitted. He nudged Micca on the side when the Flyer didn't say anything.

"Oh, all right. I'll do as you say." He stood up, swaying. Sal rose to steady him.

"You Chosen are all the same. One giant stubborn bunch." There was more amusement than rancor in the statement.

Micca shrugged, too tired to argue, and set off to show Sal the way.

By the time Torren woke them a few hours later, everything was ready, as promised.

"The wagon and gear are at Hanson's Stables close to the east gate. There'll be weapons and provisions, a strong team of horses, and backup horses as well."

Sal rose stiffly from the cot set up against a wall beside the Flyer bed on which Micca was resting. "Good, we should be able to make the border in a day if we press them."

"The twins are insisting on coming along. With Micca and me, that makes four. I've already troubled you enough, Sal. There's no reason you have to come as well. There's a chance we might run into the Black Lords at some point."

"Hah! All the more reason you need me." Sal glanced over at Micca as the latter groggily rubbed the sleep from his face. "No offense, but you Chosen aren't the most militarily skilled bunch, and those mercenaries are. You're going to need all the help you can get. Besides," he added with a wink, "I could use a little action. It's been a while."

Torren half-frowned, not wanting to endanger his friend's life but also knowing he was right. With his other three companions never tested in battle, they would be at a disadvantage. "Thanks."

He left them then to clean up, and the group met not long after in Rux's office.

"We're going to have to be circumspect about this," Torren warned. "I'm pretty sure this place is being watched, and our movements reported. It's just hard to say whether it's the emissary, the governor, some other or all of them." His gaze flitted from Sal and Micca to the twins and back. "Our arrangements for today were made through a third party, so I'm hoping this has helped make our plans unclear. To minimize trouble, the five of us should leave here in two groups and eventually meet at the stables." His gaze met the twins'. "Some gear has been prepared for you two so you can disguise yourselves as Landers after you leave here. It should make it easier for you to ditch anyone who might decide to follow you."

"How long will it take to get to the mountains and back?" This came from one of the twins.

Sal answered the question. "We can make it to the border in a day if we push hard. The mountains aren't much farther, but our pace will drop dramatically. If the directions we got from our source are good, there's a chance we might make it there around nightfall tomorrow. I'd say we could make it back by the day after as long as we don't run into any complications."

Torren turned to Rux. "Do you think you could contact Mallean and see if she could delay the council for that long? If it would help, have her divulge to the council something of what we're up to."

"Won't it make it more dangerous for you?" Rux's brow furrowed.

"We'll have to risk it. The atmosphere up there is very volatile, and there's no telling what will happen once the people are told what's going on. We still don't know why this was done in the first place or what else it is they're hoping to achieve."

"The Chosen and Caeldanage—heck, the whole empire—are like a red anthill waiting for the slightest thing to set them boiling out in a frenzy. No one may want to believe it, but it's looking more and more as if we're going to war." Sal shook his head sadly.

"With any luck, it won't get that far. Not if we can prove the Black Lords are behind all this and we get Larana back to normal."

# CHAPTER 23

THE TWINS LEFT FIRST, EACH CARRYING A BUNDLE AND a map drawn by Sal showing them how to get to his inn and from there to the stable. A short while later, Torren, Micca and Sal left as well, Micca once more in his Lander disguise. They headed together toward the governor's mansion, Sal and Torren looking for any signs they were being followed. A block from there, they split up.

Torren turned down an alley and waited several long breaths to see if he could spot someone trailing him. Not seeing anyone, he jogged through to the next street then headed in the opposite direction. After almost a half-hour with no sign of pursuit, he headed for the Hanson Stables and slipped in through the back. The others were already there.

Micca and Sal were talking quietly while the twins stood to the side, shifting back and forth, obviously un-comfortable in the unfamiliar Lander clothing.

"Did anyone see anything suspicious?" Torren asked, glad to see them.

"We think someone was following us," one of the twins answered, "but we lost him at the inn."

Sal spoke up next. "A wiry fellow stuck close to me for a few blocks, but I couldn't be sure he was actually follow-ing me."

"I didn't see anyone" was Micca's reply.

"Neither did I. Let's hope we were just being overly cautious." He led them to the other side of the stables to the rigging area. "That should be our ride over there."

A teenager and a balding man were hitching up a team

237

of four horses to a large covered wagon. Spotting them, the balding man said a few words to the boy then came to meet them, wiping his hands on his leggings.

"You must be the gentlemen I've been expecting. We're setting up the team right now as requested. We also have the extra horses you wanted. They should give you no trouble."

"Thank you." Torren removed several coins from his pouch. "I believe this is what was agreed on."

Hanson took the money, not bothering to count it. "The feed and water are in the back as well as the other provisions that arrived. Let me just check the horses one last time, and you should be set to go."

As he went to work, Torren led the way to the rear of the wagon. Pulling aside the back flap of the tarp and lowering the gate, he dug amongst the bags and pulled out a battered broad-brimmed hat and a lightweight bedraggled jacket.

"Until we get out of the city, it'd be best if you all stayed out of sight." He said it so only those with him could hear.

The three Flyers stared at the dark, cramped space uneasily. Sal grinned and smacked Torren resoundingly on the back.

"Sounds good. I can get some more sleep." He turned to face the others. "Come on, you three, I promise not to bite." A flash of white teeth showed from within his beard. "Not hard, anyway."

The twins exchanged a long look, even more uncomfortable than before. Torren shook his head, fighting off a grin.

"It's just Lander humor. You've nothing to worry about."

"You hope!" Sal clacked his teeth together, obviously way too entertained by his own sense of humor.

"He's trustworthy," Micca added earnestly.

Sal climbed into the wagon, chuckling softly after sending a hungry look at the twins. Micca climbed in behind him and the twins followed, reluctantly.

Still shaking his head as an amused chortle came from inside the wagon, Torren put up the tailgate and let the flap fall. He then went to the front of the wagon, slipping on the bedraggled jacket on the way.

The boy tied two horses to the back of the wagon. Hanson waved to Torren as he climbed up into the driver's seat.

"Everything's set. If you're ready to go, I'll open the doors."

"Great, go ahead." He picked up the reins from the loop on the driver's side and slipped the hat on. The brim was large enough to make his face hard to see and to definitely hide his telltale blond hair.

He clucked to the horses and flicked the reins as the wide stable doors opened. Waving goodbye to Hanson and his helper, he drove out onto the street and headed the team toward the east gate of the city at a sedate walk. No one paid him much attention as he left town. It didn't escape his notice, however, that the guard was doubled at the gate. He'd have to remember to ask Sal what other things had changed since he'd been gone.

What little he'd seen while making the arrangements for this trip hadn't reassured him. He felt even worse as he spotted the sea of tents set up over a large area of fallow fields outside the city—the emissary's troops Sal had mentioned. Their presence didn't bode well at all.

Torren followed the eastern Grand Highway until he crossed a smaller road heading north. There, he brought the team to a halt and stood up in his seat to look over the top of the wagon. He waited for several long minutes, making sure no one was coming in their direction. Finally satisfied, he leapt down to the ground and went around the wagon, removing the jacket and hat as he went.

"It appears we're clear." Opening the back, he heard at least one sigh of relief. "The seat up front will hold one more if one of you wants out. It might be better to get some sleep in here, though. I'll be pushing us hard."

"I'll stay, if you don't mind," Sal replied from the back. "Already worked me up a cushy spot here." He was stretched out on the floor of the wagon, his head on a bag of grain. The twins huddled on opposite sides, still eyeing him suspiciously.

Despite their continued misgivings, however, the twins opted to stay inside as well, since while out of sight they could remove the more uncomfortable parts of their Lander disguises. Micca, however, chose to put up with his and joined Torren in front. It was a slight stretch for him to get up and sit on the narrow bench, since he had to position his wings far enough apart for him to sit and yet manage to keep them covered and out of sight.

They set off; and once they'd turned onto the northern road, Torren brought the horses up from a walk to a light trot.

"Sal is quite a character." Micca glanced at him, holding on tight as the wagon hit small bumps on the rutted road.

"Yes, he's one of the good ones."

"He was amazing yesterday. He seemed to know everyone." Micca's tone held a touch of awe. "High, low—he treated everyone as if they were a lost brother, unless they got in his way. We would have never found out the things we did if it weren't for him." He sent Torren a furtive glance. "Did you promise him something for all this?"

He smiled, understanding the source of the Flyer's confusion. "No. As I've said before, not all Landers fit the mold the Chosen have been taught to expect. He's doing all this for one reason—because he is who he is, and my friend. I haven't always appreciated it as I should, but I'm very lucky to have him."

Micca slowly shook his head as if to clear it. "I must admit, before this, even working with my uncle, it hadn't really occurred to me to question what we think of them. But recently, I've had no choice but to take a closer look. The fact Landers helped you despite you not being one of their own, Aen's captors coming to love her and giving their lives for her, Sal's efforts—it speaks much of things we've forever assumed didn't exist.

"We all came from the same stock, only we don't think of the good traits as being shared but rather as something we received because of our standing with El." He shook his head again. "Why have we never tried to find out before?"

Torren shrugged. "Maybe it was just easier that way."

Micca thought that over for a long moment then said, "Maybe, after all this is over, it could change."

Torren made no response. After a while, Micca slipped into a light doze.

# CHAPTER 24

THE AFTERNOON WENT BY SWIFTLY. THEY MADE GOOD time on the northern road, and Torren continued to follow it as it took an eastern tilt and eventually joined the northern Grand Highway. Switching horses after a short rest, he pushed them to greater speeds on the even-surfaced road.

Those sitting in front were rotated as well, Sal eventually taking over the reins so Torren could catch some sleep.

In the late morning of the second day, they arrived at the town Torren and Larana had stopped at while working for the caravan, before they were attacked by the Black Lords. Handing over some coin, they traded the local blacksmith for several of his horses so theirs could be rested.

Nearing the border late in the day, Torren frowned as he spotted a thin trail of smoke rising where the trees opened before them. The only thing he knew of that lay in that direction was the border town, with its small imperial outpost.

Something about the color of the smoke didn't seem right. He urged the horses to move a little faster.

"Micca, wake up. We may be in for some trouble."

❦

The town was a burned ruin. The remains of homes and stores stared at them with broken eyes. The stone tower of the outpost was stained dark and still smoking. There was only silence as the four Flyers and one Lander stared at the devastation.

"What happened here?" Micca asked, his face pale. The smell of burned wood was almost overridden by the harsher stench of scorched flesh.

"I'd have to say a declaration of war." Sal's face was grim. "Look—Galt arrows and war axes. It won't take long before things escalate, mountains or no mountains."

"Yes," Torren admitted, "but something about this is wrong."

He didn't elucidate, instead going to the closest unburned body. Crouching, he inspected it before getting up and moving on to the next with almost clinical detachment. Sal followed him; the others were more hesitant. The twins wrapped their wings about themselves, as if to shield themselves from the carnage.

"By the gods, Torren, you're right!" Sal glanced quickly around him. "Where's all the blood?"

The twins glanced at one another in confusion, having seen small pools around a number of the bodies.

"What are you talking about? There should be more?" They sounded horrified.

"Look at this fellow here. See the cut in his neck?" Sal pointed at the man at his feet. "At the time of the cut, his heart would have still been pumping. On something like this, the blood would have sprayed out all over his attacker and the general area, yet all you see is a bit that looks to have pooled beside him after he fell. Something's not right."

The farther they went, the more grisly the deaths became, though there was still very little blood. Half-burnt bodies hung from doorways. Women with torn clothing and slit throats had been dumped in ditches. Children had had their heads bashed or been decapitated.

"Where are the people who did this?" Micca asked. "Wouldn't some of them have been killed as well?" The Flyer looked ill.

"Galts don't leave their dead behind. Their bodies are burned on pyres to be claimed anew by the First Mother." Sal rubbed absently at the scar running along his cheek. "Nomads, the whole lot of them, and secretive to boot. But that they'd be responsible for something like this..."

"No Galts actually died here today. I doubt they were ever even here."

Micca, Sal and the twins turned to stare at Torren. He emerged from the hulk of a burned home smeared with soot and grime. "There are no spots of blood showing where their bodies fell."

Sal glanced around, verifying the truth for himself. "This has been staged, then."

"Staged?" Micca's puzzlement was plain.

"Yes," Torren responded, "staged." He pointed at the body-littered street. "These people are dead and they were killed, but not by the means the evidence suggests. Some of the cuts are inconsistent with the weapons. Then there's the lack of blood, as if the people were cut after they were already dead." He pointed to the right. "If you look, you'll notice some of the guards aren't properly dressed for battle or even to be outdoors, as if they'd been caught unawares or didn't come out to fight at all."

He reached them and dropped a half-burned bag at their feet. "This grain's been poisoned. The water, too. You can tell because the reflection in it isn't quite right. These people never got the chance to fight a battle."

Silence hung for a moment as his revelations sank in.

"But won't those from the city know this?" one of the twins asked.

"It'll depend." Sal kicked a loose rock. "Many will jump to the conclusions they want despite signs to the contrary. And if it's someone inexperienced who's leading them, they won't even know enough to see the signs."

"Something similar to this is bound to happen across the mountains as well, isn't it?" Micca stared, his voice small.

"You can count on it. If they spark off both sides at once, no one will stop long enough to realize they're being manipulated." Torren grimaced. "The Black Lords have done their job well."

"Do Landers not think it wrong to murder their own kind?" Styn stared at the bodies as if the answer might be among them.

"Aye," Sal told him, "for the most part. But there are

times when it's necessary, times when it can't be helped-to protect country, family, those who've hired you to guard them and the things I've just said. This, though, was none of those." He slowly shook his head. "We've got to tell someone." His expression was grim. "We've got to stop this before it gets out of hand."

"No! We can't turn back." Micca stood with fists curled at his sides, wings splayed as if for flight. "There's no telling how long the council can hold our people back or Aen's condition remain stable."

"It might be too late if we wait," Sal countered. "I could take one of the horses and ride back. It might slow you down a little, but at least both tasks could be done."

Micca's face screwed up with indecision then cleared slightly as he nodded.

The twins stepped up before Torren could give his opinion one way or the other. "There's another possible solution. One of us could just fly back. This way Aen's rescue wouldn't be threatened in any way."

"Would one of you be willing to do this?" Torren asked.

The twins' eyes met for a moment and then Mar stepped forward. "I will. And I can take some of the grain and water back with me as proof."

"Let's do that, then. Take the proof back to the nearest station and then to Dom Rux. He will contact the governor and the emissary. The more we can wreck whatever plans the Black Lords are up to, the better."

The group quickly got what was needed and outfitted Mar with food and drink for his journey. The two brothers hugged one another warmly, and then Mar took flight back in the direction they'd come. Styn watched his brother until he disappeared from sight. Then he and Micca piled into the back of the wagon as Sal and Torren took the front, and they set off again.

It wasn't long before they began their ascent up into the mountains. Following the directions Sal had paid for in Caeldanage, they stayed in the main pass, the road turning to dirt from stone once the highway ended at the border. In the decreasing light, they kept their eyes open for the branch of the road that would take them where they wanted to go.

"There should be a camping place for caravans not too much farther up. A trail splits from there toward where the ex-priestess lives."

Torren kept watch, wishing the road left them a little less exposed to curious eyes. "Think we can make it before we lose all the light?"

Sal pondered on this a moment before replying. "It's hard to say. But I think we'd better try, or Micca will go bumbling in the dark trying to find it anyway." He flashed him a grin.

Torren could picture that eventuality only too clearly. He absently clucked to the horses so they'd pick up the pace a little.

They reached the caravan resting spot and found the trail turning off to the right. From what he could see of the area in the fading light, a caravan had been by recently. They'd met no one on the road, and no wagons had been in the burned-out town so he supposed they must be going north. He didn't envy them the sight that would probably greet them in the first village on the other side.

The new trail proved to be little more than a beaten track but was not badly overgrown, so it was being used on occasion. Their pace, however, slowed down to almost a crawl as the trees and other foliage stole what little light they had left.

"It doesn't appear we'll make it," Sal commented sadly.

Torren brought the team to a halt. "I think there's a lamp amongst the provisions. I can lead the horses."

Sal nodded and jumped off to fetch the lamp. Torren descended more slowly, stretching stiff muscles. Micca and Styn returned with Sal, the lamp already lit and doing its best to push back the deepening darkness.

"Are we almost there?" Micca stared up the road, though it was possible to see very little beyond the range of the light, stretching and retracting his wings nervously.

Torren reached for the lamp. "If the directions are accurate, it shouldn't be much farther."

Sal pulled it back. "I'll hold it. I need to stretch a bit, and you're going to have your hands full, anyway."

The two Flyers took the seat, their wings wrapped around them, as Torren and Sal moved in front of the horses. Sal went ahead, Torren leading the horses only a few steps behind him.

Two of the moons were in the night sky by the time the trail opened up into a small clearing.

"Is this the place?" Torren glanced around, seeing no evidence that anyone lived there.

"I think so," Sal replied, "though our man said it'd be hard to tell."

Micca got off the wagon. "He said the house would be on the north end, didn't he?"

"And who might you all be?"

Startled, all four turned to the left, in the direction of the unexpected voice. No one was there.

Sal took several steps in that direction, bringing light to the darkness. He stopped as the vague form of a woman appeared at the edge of the clearing. She wore a heavily patched cloak, its hood hiding her face in shadow.

"Look a strange lot, you do." Her soft cackle made the horses fidget. "Guess you're not with the others."

"Are you Mala?" Micca asked. "We've come on some urgent business to see her."

The woman chuckled. "Urgent business, is it? That's new." She stayed on the fringe of the lamplight as she moved farther into the clearing. "What would this urgent business be?"

The four of them exchanged furtive glances, knowing perfectly well she'd not verified she was the person they were looking for.

Torren decided to risk it. "We've heard of a poison made from a rare fish in these mountains. We've come in the hope that since you've sold this poison, you might also be able to sell us an antidote."

The atmosphere in the clearing filled with expectation as they waited for the woman to speak.

"From a fish, you say?" she asked quietly. "Was this poison ingested or placed into the blood?"

"We believe it was ingested. It's put a young girl into a sleep from which she will not awaken."

The woman in the robe nodded absently. "It was given to a Chosen, was it?"

"Yes! And she's been in this hideous sleep for days,"

Micca's voice exploded across the clearing. "She has to be made whole. We will give you whatever you want."

"Micca!" Torren's annoyed bark brought the Flyer up short. Micca stared at him with confusion tainted with defiance. Leave it to the Chosen not to know you never offered anyone the choice of whatever they wanted, let alone do it with the very person who had most likely sold the poison to their enemies in the first place.

"You needn't worry," the woman told Torren. "I will not take his words to heart." She laughed. For the first time, she drifted into the light, coming closer to Micca. "It's rare to see such openness. Especially in these troubled times."

Torren frowned as she brought her hands up to throw back her cowl.

Mala was fair, with a mature, handsome face and dark, penetrating eyes, nothing at all like what her gravelly voice suggested. Dark auburn hair fell to her shoulders, showing the barest hint of gray. Micca stood as if transfixed as she lightly ran her fingers over his half-retracted wings, her eyes trapping his.

"Very rare, indeed."

"Come, all of you," she said, turning away from the Flyer. "It is late, and there are those about whom I am sure you do not wish to become aware of your presence. Bring your wagon and horses as well."

She took Micca's hand and pulled him away from the light.

Sal sent Torren a questioning look, which Torren answered with a shrug. He placed one hand on the hilt of his sword as he tugged to get the horses moving again. Sal ventured forward to keep Micca and Mala within the field of light.

At the north edge of the clearing, Mala stopped, and taking a long pole from behind a tree, she poked up at something hidden beyond a twisted branch. Torren was forced to pull hard on the horses to keep them steady as the branch snapped out, making a sharp sound like a whip, and disappeared up. In the dim light, with the leaves now out of the way, they could see another, smaller clearing sheltered by heavy foliage and a small stone cottage almost totally overgrown with moss and ivy. Its doorway stood open to the night, the pale glow of firelight filtering out.

Torren saw Mala smile a secret smile as they took all this in.

"You can tether your horses in the back. There's a well there and a trough. One of you reel in the rope there to set the branch back." She pointed up one of the larger trees, her voice showing she was used to giving commands and having them obeyed. "Meanwhile, we'll make us some tea."

She squeezed Micca's hand and then let him go, her smile inviting him in. Not waiting, she stepped into the cottage. After throwing a glance back at the others, he followed her.

Gingerly, Styn drifted to Torren's side, his wings quivering. "Will he be all right with her?"

He could see the fear of Lander childhood stories in the Flyer's eyes.

"I doubt she plans to eat him. But why don't you go keep an eye on him all the same."

Styn nodded, hurrying over to the doorway.

Torren unhitched the horses and rubbed them down before getting feed from the back of the wagon and drawing water from the small well. Across the clearing, he heard Sal grunt as he strained to put the counterweighted branch back in place. As soon as they were done, they headed for the cottage. He could hear Mala's voice, though it sounded softer than before, almost as if its previous roughness were from disuse.

His first glimpse into the place showed him it was warmer and brighter than he'd expected. Colored glass bottles and jars of unknown substances took up shelves built into the walls, every one labeled in a strange angular script. Heavily carved rafters held up a tall ceiling, making the room seem larger than it was. A wide hearth hung with tied bundles of drying flowers spread light and warmth, welcoming them in.

"Balance is the First Mother's way," Mala was saying. "That's why there are plants that flower while others don't, some that have edible roots while others bear fruit. Why some can heal while others kill. There's no good or evil to these elements, only in how they are used. The very thing that can bring life to one could also bring death to another. Yet still, in all of it, there is balance." She looked up as Torren and Sal came in. "Ah, there you are. Your tea has been waiting."

She indicated two misshapen ceramic cups set close to the fire. Micca and Styn sat on stools drinking their own.

Hesitantly, wondering if it were wise, Torren retrieved a cup and passed Sal the other. He noticed Mala watching them intently from her perch. The flickering light from the fire added and took years from her face from one moment to the next.

Torren took a tentative sip of his drink, trying not to give credence to the small voice in his head warning him it could be poisoned. Warm tea tickled his tongue, a tart, fruity aftertaste mingling with it as the scent of cinnamon wafted to his nose.

"This is good!" Sal grinned with pleasure.

"You seem surprised it's not poisoned." Mala smiled at Torren's sharp glance. Styn looked away guiltily. It seemed a conversation about this had occurred before.

No one said anything for several minutes as they drank.

"You took a great risk coming here," Mala told them. "Things are becoming unstable in these parts."

"How do you mean?"

She glanced over at Micca. "Dark forces are at work, trying to bring the world into chaos. Order normally holds sway with chaos beneath it, maintaining a balance, as it should. But now chaos is seeking to overthrow order and at a higher magnitude than ever before."

Torren felt goose bumps rise on his arms as her eyes took on a strange, unfocused look.

"It will sweep us all away if we're not careful." Her voice changed to a faraway tone, as if she were no longer entirely with them. "Many will die, and the cancer will eat away at all until everything we know is destroyed."

"Give us the cure for the poison, and perhaps we can stop it before it goes too far." Her portents reminded Torren only too well of his own dreams of late.

"Is that so?" Mala's intent, dark eyes latched on to him,

her smile almost feral. He had to struggle not to look away. What was she playing at?

"There is a cure, but I will have to make it. You will all stay here the night, and by dawn I will have it ready for you." Her gaze didn't leave his face.

"And what will you demand as payment?" Torren tensed, sure he wouldn't like the answer.

The feral smile returned. "Well, that is the question, isn't it?" Her eyes gleamed. "What if I fancied a pair of wings for myself?"

Micca blinked in confusion. "How would we give you that? Only El has the power to grant wings."

Her eyes remained locked on Torren as she answered. "There are other ways, are there not?"

Torren's face grew hard. He'd heard some of those who dabbled in magic used the Flyers' feathers for other things than stuffing pillows, sure that magic resided in the wings as in the Chosen themselves. Did she think she could use them to become one of them?

Micca rose from his stool, looking from one to the other, then paled as Mala's meaning slowly dawned on him. "If...If it's what you require, so be it. Take mine." His wings quivered at the ends.

Torren growled a warning. "Micca..."

The young Flyer turned on him. "No, I mean it! I'll do anything for Aen, for El. And if you can survive their loss, so can I." His eyes looked wild.

Torren stepped toward him, not knowing what he intended to say as his mind screamed that this young fool possessed no idea of the consequences of what he was willing to give away.

Loud laughter stopped him in his tracks. It was sweet and joyous.

"How brave you are!" Mala took a surprised Micca's arm and flashed a grin in Torren's direction. "Don't worry. I don't want any wings. Money, a few feathers and promises for rare herbs from places only the Chosen can reach will be more than enough to satisfy me."

Micca reached behind him and yanked several from his wing, grimacing. He extended them toward her, drops of blood gleaming on the ends. "You have my word."

Mala's features changed to resemble those of a child, her eyes bright. "Wait for me here. I will not disappoint you."

She released his arm then cut past the others to the door. Without another word, she was gone; and soon it seemed as if she'd never been.

"Are other Landers like that?" Styn wondered aloud.

"No," Sal responded, his answer just as quiet as the question. "She's a weird one in our books as well."

Torren couldn't have agreed more. "Since it appears we have a long wait, we should set up a watch and the remainder get some rest." The others nodded. "I'd also suggest we not sleep in here."

He got no argument on that, either.

Sal volunteered to take the first watch. Micca and Styn retired to the inside of the wagon while Torren took out a blanket and settled down not far from the where the horses were resting. Each of them took a turn during the night, its quiet oppressive but revealing no signs of danger or of the ex-priestess.

Close to dawn, Torren was awake, and not sure as to the reason. Sitting up slowly, listening for what might have roused him, he spotted Styn pacing in the clearing not far away, his wings coiled about him.

He scanned the rest of the clearing and ended up at the cottage. He blinked as he saw the door was open.

Rising to his feet, he approached the pacing Flyer. "Did you see her?"

Styn frowned. "See who?"

Torren pointed at the open door. Styn's eyes widened. "Get the others."

Nodding, Styn hastened back to the wagon. Torren headed for the open doorway.

Peeking inside, not sure what to expect, he found the fire rekindled with fresh wood and a large pot set over it.

"Ah, there you are."

He almost jumped at the voice as a dark shadow disengaged from the wall on the left to become Mala.

"I've put breakfast on—it should be ready before long. I doubt it would be prudent for you to leave before the sun arrives." She took a bowl of dried berries and dumped them into the pot.

Torren tried hard to find his voice. "Were you able to get what you needed?" He felt cold and nervous all at once.

"Oh, you mean this?" She pulled a small vial from the folds of her cloak. A dark-green liquid swirled inside it. He was forced to control the urge to step forward and snatch it from her hands.

As if she knew his thoughts, Mala laughed like a small child and set the vial on the table closest to him.

"Go ahead, take it." Her eyes sparkled. "Make sure to only give her a few drops. Too many will not improve her condition."

He picked the vial up, not sure if he actually believed it was what it was.

"Torren, did she...?" Micca popped in the doorway, his eyes immediately drawn toward the vial as he spoke. "Is that it?"

Sal and Styn crowded behind him.

He nodded and placed the vial carefully into a small pouch on his belt.

"No, you cannot leave until after breakfast."

Micca closed his mouth as she snatched the question from him before he was able to ask it.

"You were lucky last night, but if you try to leave as you came, they will definitely find you."

The others stepped inside the cottage.

"Have you seen them?" Sal asked.

"Oh, yes. But they did not see me." Mala grinned, a devious look in her eyes.

"Where are they?"

She turned to stir the contents of the pot. The scent of fruit and porridge filled the room. "Scattered through the pass in small groups. I believe they plan to start making trouble on the caravans and anyone else they find there."

She slowly shook her head, as if thinking the whole idea foolish.

Not long after, she scooped hot porridge into shallow wooden bowls and passed them around. "Eat up, it'll give you energy. Unless you still think I'm out to poison you?"

Styn's cheeks colored, he having eyed the unusual Lander fare with misgivings. He set to eating like the others. Mala chuckled softly.

By the time they were done, the sky had started to lighten. Torren pulled out the money he'd brought for the antidote. Mala gestured for him to put it on a shelf, as if it held little importance to her.

Micca moved to stand before her. "I gave you my word and I will keep it. You shall have what else we promised as soon as Aen is made well."

Mala smiled. "I am not worried. I know you will."

Drifting outside, Torren and Sal soon got the horses hitched. Sal and Styn climbed into the back; Micca settled up front. Torren held the reins to hold the horses steady, knowing what would come next.

Once they were set, Mala released the branch hiding the clearing from view to open up the way for them. She grabbed Torren's arm as he made to climb up onto the wagon.

"Fix what must be fixed." Quicker than he thought possible, she reached up and caressed his cheek. "Make the Vassals whole."

Confused, but trying not to show it, he climbed on up. Mala's sharp gaze followed them as they set off.

# CHAPTER 25

S HADOWS STILL CLUNG TO THE SIDES OF THE ROAD BUT were clearing by the moment. Though they could see, Torren still didn't dare allow the horses to move at any great speed due to the rough road. Mala's last words still rang through his mind, making him feel uneasy. What had she meant by "Vassals?"

Micca sat forward, only too eager to be gone. "It's almost over. Once we get the antidote back to her..."

Torren nodded, hoping it would be as easy as that. But even if they saved Larana, there would still be the matter of finding out who'd poisoned her, and the continuing problem of the Black Lords.

Keeping the ex-priestess's warning in mind, he searched for signs of trouble. Something flashed up ahead in the trees, catching his attention. They had almost reached the place where he thought he'd seen the strange glint of light when he realized the canopy above them wasn't made totally of leaves.

His stomach knotted painfully as he realized he'd seen something resembling this before. "Sal!"

Dropping the reins, he shoved Micca off the side of the wagon as a faint twang sounded from overhead. He leapt from his seat as something crashed down. Rolling as he hit the ground, he avoided most of the net as it enveloped the wagon.

The horses whinnied and stamped in panic but couldn't run. He grimaced as he got to his feet, one of the weights attached to the net having hit him on the thigh.

A twig snapped off to the side; and Torren instantly

dropped into a half-crouch, pulling out his sword as he turned. Anger overrode fear as he spotted a man in dark leather. They were trying to do it to him again. This time they wouldn't find his group so obliging.

A snarl on his face, he dove at the mercenary, slashing at his legs. The man backpedaled, obviously taken aback at the ferocity of the attack.

"Marin!" The mercenary stumbled and fell.

Torren didn't hesitate. He slashed down at the man's arm, cutting through the tendons. His second thrust ended the mercenary's screams as he drove the sharp blade into the man's throat. *For my father!*

A cry of pain spun his attention back toward the wagon. Yanking his blade from the corpse, he hurried to look for the others. He spotted Micca bobbing in the air close to the back of the wagon, a thin trail of blood running down his arm.

"Bastards." Fury infusing him to the core, Torren hobbled toward his companion as Micca dived down at the man standing below him.

"Get away from them!" Micca avoided the mercenary's sword as it swung toward him, soaring up out of range.

"Come on down and make me." The mercenary flashed an obscene gesture with his free hand while he slapped at the netting with his blade to keep Sal and Styn from getting out of the wagon. "Drog, get on over here!"

Torren realized the mercenary assumed the previous cry had been his. Soon, he would show the man his error.

"Let me out of here, you coward. I'll give you the entertainment you're looking for." Sal returned the rude gesture twofold.

"You will now die."

The mercenary half-turned, and saw Torren. The cruel amusement on his features flushed away as he met Torren's murderous gaze. Micca made another dive; the man slashed at him again to keep him back. Then, after a last glance between his two opponents, the mercenary abruptly turned tail and ran.

Torren started after him, sending shooting pains through his thigh as he abused it further, but was unable to keep his target from getting away. Gritting his teeth, he bent down and, in one fluid move, removed the knife from his boot. He threw, aiming for the unarmored head.

The handle, instead of the blade, smacked into the back of the mercenary's skull and sent him sprawling. Torren was on him before he could recover, thirsty for blood. He brought his blade up to finish the man, but at the last instant turned the blade and pummeled him on the head with the hilt instead.

The man's body went limp. Torren tried not to think about whether he was doing the right thing or not.

Sal emerged from the trees, his naked blade before him. "Torren!"

"Over here." He rose shakily to his feet, his thigh throbbing. The bruise wouldn't be better for all this exertion.

"Is he dead?" Sal looked at the body on the ground.

"No. Figured he might have some stories to tell us."

Sal grinned. "Yes, this fellow might become very popular back in the city." He knelt, checking the large bumps on the mercenary's head. "That'll smart later. Serves him right."

His grin grew. Sheathing his sword, he grabbed the unconscious man and lifted him up over his shoulder like a sack of grain. "Won't be safe to linger around here."

"Right. No telling how much more trouble they'll decide to make for us." Torren felt tired, his fury gone. He hobbled behind Sal back to the wagon, where they found Styn bandaging Micca's arm.

"How bad is it?" Sal asked.

"It's just a nick," Micca told them, looking embarrassed. "He caught me by surprise."

"We're damn lucky they missed us last night." Sal dumped the mercenary on the ground and, with Styn's help, lifted the net enough to slip inside the wagon to retrieve some rope. "I don't think things would have gone this well."

"Let's not give them an opportunity to try it again, then," Torren said as he helped Sal tie up their guest. "Styn, if you're willing, I want you to fly a little ahead and make sure there are no more overhead surprises waiting for us."

Styn nodded. "Of course."

As soon as they finished trussing up their prisoner and stowing him securely inside, Sal, Styn and Micca removed the netting from the wagon while Torren calmed the horses as best he could. Their eyes rolled, and sweat covered their coats—Larana's uncanny skill with animals would have been very useful here.

By the time the nets were removed, the horses calm and everyone had taken a bit of rest after the heavy labor, a large part of the morning had passed. Sal gave Styn his sword before the Flyer lifted into the air to scout ahead, where he discovered three more nets but no one manning them.

As soon as they reached the main road and the area where the trees thinned, he returned to the wagon, and Torren set the horses to as fast a pace as they could manage.

They donned their disguises again. All eyes were busy keeping watch as they followed the Grand Highway south.

The fires at the border town were no longer smoldering, but the reek of decaying and burning flesh was worse than before. Soldiers filled the streets, cleaning up the main road and piling bodies onto a pyre at the edge of town.

Though Torren expected to be stopped and questioned, a captain waved them on. Mar obviously had gotten their message through, at least to the closest of the military stations.

They made good time, once again rotating the horses at rest stops. They traveled all day and through the night; Caeldanage came into view late the following afternoon.

Other wagons and horses were lined up on the road waiting to get into the city; each was inspected thoroughly before being allowed inside. Micca stared at the line of farmers and merchants before them, occasionally glancing back the way they'd come as if still expecting pursuit.

Torren felt less at ease about what they might encounter inside the city, before they could make it to the embassy. The city had sheltered members of the Black Lords before.

"Torren, give me the vial. I will take it to her now."

Micca turned to him, his face tight. "Every minute we sit here will only give our enemies more time to try something else."

Torren stared at him, thinking himself a fool for not having thought of that. "Do it. I'll find Dom Rux and fill him in as soon as we can get inside the city." He removed the vial from his pouch and handed it over as Micca removed his cloak. "Just a few drops, don't forget."

Micca accepted the vial as if it were more fragile than it seemed and tucked it away. "I won't."

Torren realized that Micca's taking the antidote to Larana would not only speed her recovery but would save him from awkward goodbyes. Still, a part of him was saddened at not being able to see her smile one last time.

Micca's eyes met his. "I will get this to her, I swear it." He suddenly grinned. "I will thank El for the rest of my days you were here to help us through this."

Without further ado, he stood up on the seat, tossed aside his cloak, spread out his wings and took flight. Torren watched as he headed upward toward the floating island. His attention was drawn back to the city gates, however, as the astonished shouts of those before them brought a number of soldiers running in their direction.

"That's done it." Sal came to the front of the wagon and climbed onto the bench. "I take it Micca could wait no more?"

Torren nodded, keeping his eye on the approaching soldiers.

"This could be awkward."

He agreed but still didn't regret his decision.

Neither man made any aggressive moves as the wagon was quietly surrounded. A burly captain stared them up and down.

"The Flyer we just saw, was he with you?" His tone left little doubt he already knew the answer.

Torren removed his hat, revealing his light-colored hair. "We need an escort into the city. If possible, it should take us all the way to the Chosen's embassy."

The captain raised a brow. "I have orders to do just that for a lot fitting your description."

"Well, then, man," Sal proclaimed, "what are you waiting for?"

Blinking twice, the captain turned away and commanded his men to clear the way. Grinning, Sal poked
Torren in the side.

"If I'd known this is what it'd get us, I would have sent
Micca flying a might sooner."

With an escort on either side, they hurried through the
streets. By the time they reached the embassy, they were
surrounded by even more guards, these with the livery of
the emperor. They made way as Styn, Sal and Torren got
off the wagon to go through the gate. Within were even
more guards, these belonging to both the emperor and the
Chosen. Torren and the others were immediately shown
inside.

"You're here!" Rux rushed forward as they came in. He
gave heartfelt hugs to each in turn, including, much to
everyone's surprise, Sal. In his common room, a large table had been set, papers strewn across it and several large
chairs placed around it. From one of them, the governor
gave them a brief wave. Von Duren rose from another, his
face impassive, and Mar from a third, his tired expression
brightening at the sight of his brother and the others.

Rux did a visual inventory, marking Torren's slight
limp and the absence of the last member of their party.
"Where's Micca?"

"He's gone on ahead to deliver the potion." Torren
glanced at the two Landers in the back, wondering if
they'd been told any more than before.

"That's wonderful news." Rux's wings flexed with excitement. "As you can see, the intelligence you sent back
was received." He signaled for them to come further into
the room.

"A contingent was sent before sunrise to the border
town to reinforce those already there from the outpost."
Von Duren half-nodded as they came closer. "More men
will be leaving presently to secure the pass. We're now
currently discussing strategy on how the Chosen might
intercede with Galt on our behalf to explain what has
been happening. This could have all gone horribly wrong

if not for your message. Some didn't want to believe it as it was."

"Your men will need to be careful in the pass," Torren advised him, ignoring the credit he was being given. "Though the Black Lords look to be spread thin, they've set traps in the trees and possibly concocted other impediments. We arrived late, so they missed us till morning, but any kind of large force won't be able to slip by."

Von Duren took the information in stride. "We've obviously underestimated this group for some time. It won't happen again."

The doors were thrown open. As they all turned around to stare, seven armed Chosen strode into the room.

"There he is!"

Before anyone could ask what was going on, six of the guards surrounded Torren and jostled him away from Rux toward the wall. He didn't resist, as puzzled as all the rest.

"Stop this at once!" The ambassador's face was livid. "What is the meaning of this?"

The seventh guard, dressed in golden armor, stepped forward and gave Rux a half-bow. "I apologize for the abrupt interruption, Dom Rux, but we feared for your safety."

The ambassador exchanged confused glances with Torren as the leader cast dubious glances at the rest of his guests.

"Explain yourself."

The golden-armored Flyer removed a rolled parchment from a concealed pouch. "Orders have been issued by the council for this person's arrest."

He extended the writ; and though Rux took it, he made no move to read it.

"Arrest? Torren? What are you talking about?" His incredulity was palpable. "Lar's son is a hero to our people."

Torren frowned, suddenly sure they'd been somehow outmaneuvered. The Flyer's next words only served to confirm his suspicions. One of the squad's spears painfully jabbed his side.

"This man is not a hero. He's committed the gravest sin

that could be committed against our people. He ended the life of one of our own."

Shocked stares were exchanged all around, though Von Duren, the governor and Sal were still puzzled.

"That is a vile accusation." Rux glared at the armed men. "Who has he supposedly killed?"

The Flyer looked away. "It is not for me to say. My orders are explicit. He is to be apprehended and brought to the council for judgment." He glanced over toward his men. "Secure him."

The guards pressed closer to Torren as one of them quickly divested him of his weapons. As soon as he was done, two others grabbed Torren and twisted him around, roughly shoving him against the wall and seizing his hands to tie them behind his back. He grunted in discomfort but did nothing to resist.

"Enough." Sal stepped forward, the sound of metal sliding out of leather loud in the room. "I don't care why you're doing this, but I'm not standing by while you feathered fools mistreat my friend."

Several of the guards turned to face the Lander.

"Sal, don't." Torren could only see him out of the corner of his eye, saw him stop at his command. "It'll only make things worse."

Though his friend meant well, Torren didn't want to become the reason for a Lander/Chosen war, not after all he'd been through to stop one.

Sal didn't back down. He glared at the guards, who returned the look and took firmer grips on their weapons.

"Your friend is right, listen to him." Von Duren had his eyes on the armed men, his face taut.

Rux stepped over to Sal, placing his hand on the Lander's tense shoulder.

"This man is correct, and it is to our shame he has to point it out to us." His gaze slashed across the guards and focused on their leader. "Whether the charge brought against the son of Lar is true or not is irrelevant just now. Torren is and has always been a Chosen, and you will treat him with respect, not like some animal."

The commander met Rux's stern gaze but eventually

looked away, his wings folded back. "My apologies, Dom Rux. It should be as you say."

The guards backed a step away from Torren, releasing him.

"You will have to remain bound. My orders were explicit on that account." His voice gave no hint he was sorry for this. "We will extend you every courtesy as long as you do not resist."

Torren turned around and nodded.

"Commander, I will be coming with you," Rux said. "A grievous error has been committed here, and it will be corrected."

"As you wish." The commander's expression conveyed how unlikely he thought it was an error had been made.

"We will come as well," the twins said as one, moving to stand on either side of the ambassador.

Von Duren spoke, his words laced with command. "Ambassador Rux, do not hesitate to call upon us if we can help in this matter."

A number of things seemed to be implied in his words.

"Thank you for your support," Rux told him. "But I am sure it won't go that far."

The commander looked unhappy at the exchange. "We should go. The council expects our return."

He signaled to the other guards, who motioned for Torren to proceed.

"There's one thing I'd like to know before we leave here," Torren said

The commander hesitated, half-turning to look at him. "What is it?"

"Has any news on Aen been released?"

The commander scowled at him, letting him know he wasn't privy to Larana's true condition. "You have no need to worry about the Vassal. You'd be best served to spend your time worrying about yourself."

He left the room.

Torren threw Rux a knowing glance as the guards prodded him forward. Something must have happened to Micca. He hadn't made it through. Torren sighed, realizing they'd been outmaneuvered in more than one way.

Rather than go out onto the lawn, the commander led

the group upstairs and onto the roof. Aen's Wings sat prominently in one corner but was ignored.

"You, take him." The commander signaled to four of his men and Torren was swept off his feet. Two of the Flyers took his legs while two others hooked their arms through his.

"Wait!" Styn stepped forward. "Place him in the Wings. We'll take him up. It would be less dangerous."

"And allow him to soil it more than he already has? No. His Lander contamination will not be spread to Aen by me." The commander's face was set.

Contamination—so that's what it had finally come down to. Torren was astonished to discover he was angry. But it was what he'd originally expected, wasn't it? Why should it matter that it'd taken so long to happen?

Helpless, wholly at the mercy of his captors, he was carried into the air. He tried hard not to look down as the city shrank below him—this was far worse than when Micca had taken him. The four guards struggled to rise in unison but had a hard time of it, giving him the impression he could be dropped at any moment. If they did, he would die. It only made him feel marginally better when he spotted Rux and the twins flying close beneath him, just in case.

The sun sank below the horizon, its light reminiscent of spilled blood, as if heralding things to come.

# CHAPTER 26

THEY LANDED CLOSE TO THE WESTERN EDGE OF THE floating island; Torren was very glad to have solid ground once more beneath his feet. Surrounded, he was led deeper into the island. Curious bystanders watched the strange group as they filed past, but no one got too close.

They seemed to be headed toward the coliseum and, he assumed, the council chambers beyond. As they drew closer, however, the commander's path didn't deviate, heading directly to the coliseum. Once inside, he turned down a smaller hallway.

Ten armed Flyers stood along the corridor, staring at them curiously as they passed, as if they weren't aware of exactly what was going on. A table had been propped against one of the storage room arches to create a make-shift door. Two guards stepped forward to move the table aside only long enough for him to be shoved through.

"Torren!"

He looked up and found Micca sitting on top of a bale of cloth. The Flyer's features wavered between joy and sadness. He quickly leapt down from his perch and untied the ropes binding Torren's hands.

"Micca, what are you doing here?" There was something else Torren wanted to ask but wasn't ready for the answer that would surely follow. Larana's hair clip rested heavily against his chest.

Micca let the rope fall to the floor. He would no longer look at him.

"They haven't really said, only that it was on the council's orders." He looked up, his face filled with torment. "I was almost there. I'd almost made it." He turned away, fists clenched at his sides, his wings drooping. "I'd just reached Aen's home when Tyleen rushed out and stopped me from going inside. She was upset. She told me guards had been posted in the house and that they were waiting for me and for you. She was talking so fast I could make out little of what she was saying. She grabbed me and insisted I needed to get away from there.

"I went on to Tel Mallean's, hoping she would explain what was going on, but before she got a chance to, guards flooded the house. Chosen I know, Chosen I've worked with, treated me as less than nothing and dragged me here. No explanation, no apology. One of them even called me unclean..."

Torren waited for several heartbeats for him to go on, but he didn't.

"So, they have it, then." It had all been for nothing.

Micca turned to face him, his face pale. "What?"

"The vial. They took it from you. Our enemy has won." He sat down on one of the smaller crates feeling deflated.

The Flyer's eyes brightened. "No, they haven't won quite yet."

Torren looked up at him.

He lowered his voice to a whisper. "They might have gotten me, but not the vial. I was able to slip it to Tyleen before they took me." Some of the sadness returned to his eyes. "Though I wasn't able to tell her what it was or how to use it."

Torren felt more relieved than he would have thought possible. There was still hope. "It's not over, then. Good. Now we just have to get out of this."

"Did they tell you anything?" Micca asked.

"Not much." He got up and started to pace the small, cleared area in the room. "Because Dom Rux was there, though, I was able to find out more than you did. It seems I'm wanted for murder."

"*What?*" The Flyer blanched. "How could they accuse you of such a thing? Who have you supposedly killed?"

Torren shook his head slowly. "They wouldn't say."

"They couldn't just fabricate a story. The council would

never fall for it." Micca's eyes were wide. "So, it means someone's been killed."

Torren didn't add that if it had happened here, as he suspected, it had to have been perpetrated by a Chosen. How much easier it would be to believe it was done by a wingless, tainted Flyer than by one of their own.

Micca leapt back up onto his previous perch, frowning. "This is very bad..."

"Yes."

Time passed and Torren paced, wondering when their unknown foe would make a move. Thinking over what few options he might have, he eventually remembered the papers. Since Sal had handed them over, there'd been no chance to really study them.

Taking them out, he glanced through them, hoping what he needed could be found in there. His life, Micca's and even Larana's might depend on whatever he could glean from them.

He was almost through looking when the answer came to him. "Micca—"

He stopped as the table was pulled away from the doorway. Micca leapt down in front of him, using his wings as a screen so Torren could quickly fold the papers and put them away.

The Flyer in golden armor who'd brought Torren in filled the doorway.

"Vil?" Micca took a step forward. "Won't you tell us what's going on? Why are we being accused like this?"

The cold look the commander gave the young Flyer brought him up short. "It is not my place to tell you anything."

"But, Vil, you know me. I'm a loyal servant of El. Why are you treating me this way?"

"You will allow yourselves to be bound and then taken to face the council." Vil didn't look at Micca directly but rather a point past his shoulder. "They will inform you of the details of your crimes."

He moved out of the way and signaled for four guards to come inside.

Torren submitted to having his hands retied, though this time they did so in front rather than the back. It was obvious Vil and possibly a number of the others thought them guilty. But whom were they supposed to have killed? How had the murder been connected to them? The fact they believed he'd done it didn't surprise him. Just as in Larana's original disappearance, it was easier to blame an outsider than face that the deed had been done by one of their own. But Micca?

Once they were bound, long tethers were hooked onto the ropes. Micca and Torren were surrounded by guards as they proceeded down the hallway to the main thoroughfare then into the coliseum and out the other side, close to the council's meeting chamber.

The council chamber doors were closed when they arrived. Two guards split from the group to open them while another tugged on Micca's and Torren's tethers to prompt them inside. The cacophony of voices within abruptly turned to uncomfortable silence as they were revealed.

Torren didn't look around as they were taken into the center of the wide room. He could already feel the stares, the doubts, the hatred, and didn't want to see them directly. Only when they stopped did he search for the one face that meant anything to him in this crowd.

His mother looked worn, her tear-streaked face wan. Her eyes met his, and only then did life seem to sparkle in them. She tried her best to smile, to reassure him, but fell far short of the mark. He felt a spark of anger. How dare they put her through this?

"Tel Valerian, the prisoners have been delivered as requested." Vil half-bowed to the councilor.

Valerian nodded his thanks, his expression grave. Micca's and Torren's tethers were tied to the leg of a heavy bench that had been set in the center of the room. The guards then split up to take places at the far ends of the chamber.

"Councilors, we've a most grievous duty this day," Valerian began. "We must, for the provenance of our people, decide the fate of these two wayward souls."

"You've already decided our guilt, then."

Valerian turned on Torren.

"Silence!" His voice boomed across the room. "You have not been given leave to speak."

He met the councilor's imperious stare and didn't back down. "*Will* we be given leave to speak? You look to have made up your minds regarding us already."

"It is not necessary for you to be here during this hearing," Valerian countered. "It was purely out of courtesy this was allowed."

He signaled for the guards to return.

"But why?" Micca jumped to his feet. "We don't even know who's been killed." He stared out at the council. "Why are you doing this? All we've been trying to do is help Aen."

To everyone's astonishment, Valerian rushed forward and slapped Micca's face. The Flyer went down, not prepared for the blow, as Torren leapt to get between them.

"Do not dare speak of her, filth."

"Valerian!" Mallean was on her feet. "How dare you? We didn't gather here to do our kind violence, only to try and ascertain their innocence or guilt." She brought herself under control with effort. "You have shouldered much for us of late, and we are grateful. But perhaps it is time this duty would be best served by another."

Valerian ducked his head at her rebuke; and though Torren couldn't see his face, he could read the Flyer's rage in every line of his body.

When he finally did look up, Valerian was deceptively composed. "I appreciate your concern, Tel Mallean, but though I admit I did forget myself for a moment, I promise you it will not happen again."

He stepped away from the prisoners and waved the guards back. "The prisoners will remain."

Torren helped Micca to sit on the bench. The young Flyer's face was still filled with shock, his cheek bruising from the blow. Torren sat down as well, his gaze not leaving Valerian.

The councilor made an offhanded gesture toward one of the others. "Tell them why they are here."

The stout man Valerian called upon stood, not looking in their direction. "You have been brought here to determine your innocence or guilt in the foul murder of one of our own. Yesterday, in the late evening, the...the body of Elon was found in his home. He had been stabbed."

"Elon?" Micca's dumbfounded surprise was clear for all to see. Torren felt his anger rise, able to see only too clearly the means that would be used to bring him to an end.

"Yes, Elon," Valerian said. "The two of you stand accused of being involved in his death, though only one of you of actually committing it." His cold gaze met Torren's.

"You're wrong," Micca called out, once more rising to his feet. "He's one of us. He wouldn't do such a thing."

Valerian glared the young Flyer down. "The facts will speak for themselves."

Torren felt a cold hand steal over his heart, sure of what would come next. Valerian motioned for the other councilor to proceed.

"After the...body was discovered, Elon's closest friends and acquaintances were questioned by those empowered by the council. During the inquiry, it came to light Elon and three others met with one of the accused a few nights before, and it wasn't an amicable encounter."

"It was a meeting that resulted in violence, violence perpetrated by you." Valerian pointed at Torren, his eyes alight. "I myself saw the result of your Lander violence when Elon called on me to explain his absence from the council meeting a few days ago. His face was bruised, but he explained it away as an accident. I see now I should have investigated it further. Especially since he wasn't seen thereafter, except when it was already too late."

His regret was palpable.

"Tel Valerian, what precipitated the violence Elon's friends speak of? How are you so sure it was caused by Torren?" A number of the others nodded in approval of Mallean's question.

A woman stood up in the back, her eyes filled with fanatical hate. "How relevant can it be? He's been tainted by grubs. It's obvious just by looking at him. His Lander clothes, the fact he would not return to the Chosen or his family..."

Torren cringed inside, this being the first time his reluctance to return had been spoken of aloud.

"What more reason would he need?" Shocked as well as some approving voices rose at her words. "And now he's tainted one of us." She pointed at Micca. "Who knows what he's done to Aen."

"Enough." Zelene shot to her feet, anger and sorrow on her face. "Endless paranoia will not get us anywhere. El has a reason for all that has happened, and one day He will give us the wisdom to see it. But for now, we should stick to the facts and not conjecture. All I care about is to discover whether my son is being accused of a heinous crime he didn't commit." She slumped back into her seat, as if the words drained her of energy.

Torren felt his anger burn hotter, though he knew it wouldn't help his case.

"To answer Tel Mallean's question, it's been learned Elon and the others met with the accused to do a service for El," Valerian continued. "They felt doubts as to the accused's legitimacy as a Chosen and, as servants of El, deemed it needed to be verified."

Zelene lowered her head, but her words carried plainly through the room. "What you're saying is Elon and the others didn't believe I would know my own son."

Icos spoke up then, his lined face looking more so than usual. "Elon was young and impetuous. The young don't believe their elders on most of what they say. You shouldn't take offense."

Zelene nodded but still didn't look up. Mallean, however, stood.

"And how did these young Chosen ask these questions, if Torren's own account and that of his mother weren't believed? What would satisfy them, if words had already proved not to be enough?"

Valerian's brow rose. "Why, it is my understanding they asked to see his scars. What more proof of his true lineage would there be than those?"

"And did your investigations determine whether Torren let them see them willingly? Or did they take it upon themselves to find out the truth regardless of his wishes?" Mallean's scowl showed she already suspected the answer.

"That wasn't made clear."

"Then let's ask Torren to give his account on the matter."

Valerian looked away from her steady gaze. "Even if it

were shown the others precipitated the violence, it wouldn't help him. It would only indicate he had that much more reason to come back when Elon was alone and defenseless to exact his revenge." He sent his gaze up into the waiting council. "No one in this city, in our nation, would kill one of our own. Only a stranger, one who hadn't lived with our ways, would do such a thing. Only one raised as a Lander would feel the need for vengeance. Only one raised by them would have the mind to murder.

"And he has killed before." Valerian pointed at Torren. "He has spilled Lander blood, done it for money. How little would it take, then, for him to go one step further and kill one of his own? He must be banished. His taint must be removed from us."

Torren stood up slowly but said nothing, waiting, though he knew, if he could help it, Valerian would never give him permission to speak. Others might, however.

"He wishes to speak!" Symeas spoke up when Valerian pointedly didn't look in Torren's direction.

"He will only spin tales to confuse us. What use would it be to listen to his words?" Valerian made this sound perfectly reasonable. A number of heads bobbed in agreement.

Micca shot to his feet, his wings bunched tightly against his back.

"I am not a stranger, I wasn't raised by Landers. I have lived and served El all my life. Will I be denied the right to speak as well?" His gaze was intent. "Will you gag me now and get rid of me only because I dare speak?" He turned his attention to the council, his eyes welling with tears. "How have we come to this? What have we become? Torren has given all he has on our behalf, and this is how we thank him?"

"The question more rightly should be what have *you* become?" Valerian turned his back on him. "Look at him. Do you see how he is dressed? How he is acting? Within only a short time, he's already become one of those barbarians."

"You're wrong! You're just twisting things to suit what you want to see." Micca's wings flared out behind him.

Torren still said nothing, waiting to see what Valerian would do next.

The councilor signaled for the guards. "You are disrupting these proceedings. You will be removed."

"You must listen to me," Micca turned to appeal to the council. "You must listen to Torren! The truth is being hidden from you."

A guard reached for Micca's arm, but the Flyer ducked past him, stretching his tether as far as it would go. "We are innocent!"

Torren tripped a second guard as he reached for Micca from behind.

More guards hurried onto the floor.

"Gag them."

"Stop this! Please, stop this." Zelene rose from her seat, looking ready to throw herself between the guards and her son.

"Valerian!" Mallean hurried across the floor as enough guards arrived to be able to force Micca and Torren to sit then place the gags as ordered. "We must listen. If our hearts are pure, even if they are tainted, they cannot corrupt us. As a council, we cannot make a decision until all sides are heard. Such is the wisdom El handed down to us."

Though shorter than the council's leader, her conviction made her seem taller. Valerian's left eye suddenly twitched.

"I stand with Mallean," Mides called out from his place in the third row. "Lar's son should be allowed to speak."

More and more of the councilors rose with the same demand, though there were those with dissenting voices. Icos's staff struck blow after blow on the floor, beating in time with the demand to allow Torren to speak.

"You were granted this post by the consent of the rest of us, Tel Valerian." Torren could barely hear Mallean's words over the noise of the crowd. "And you will not be allowed to abuse it. We *will* hear the truth."

Valerian's wings drooped, his eye twitching once more. His gaze dark, he turned to the council, his hands raised for silence. "They will be allowed to speak."

Zelene rushed to her son and removed the gag from his

mouth, batting one of the guards aside with her wings. Torren flexed his mouth, licking dry lips, then looked at his mother.

"I'm sorry about all this."

She shook her head, tears rising to her eyes. "It's not your fault. This is cruel, after all you've already suffered."

He stood, grateful for her support; and Zelene stayed at his side holding his arm. Mallean moved to stand with him as well, after making sure the gag was removed from Micca's mouth.

Staring once about the room at the sea of hateful, hopeful, disillusioned faces, Torren spoke.

"Micca and I are both innocent. I didn't kill Elon and neither did he. But the one guilty for this crime does stand amongst you."

"See? It already begins." Valerian pointed an accusing finger in his direction. "He proclaims his innocence without proof and then with clever words attempts to throw us off by implicating the impossible."

"If it was such an impossibility, then it will avail him nothing," Mallean retorted. "Let him continue."

Not as many voices seconded her as before.

"It wasn't my intention to come here," Torren continued. "Once I delivered Aen to Dom Rux, I thought my duty done." He felt his mother stiffen at his side, but he didn't dare look at her. "However, when Aen was stricken, Micca found me and persuaded me. He did this because some of you had come to suspect the Vassal's illness might have been induced."

"Induced? What do you mean?" There was a half-pinched look on Symeas's face.

"What I mean is someone purposely poisoned the Vassal."

The room erupted with raised voices, some angry, others totally perplexed. He waited for the torrent to die down as he felt Valerian's gaze burning into him.

"What you are implying is total lunacy," Valerian said. "Just more vain attempts to confuse us so we will forget your guilt."

"That's not true," Micca interjected. "And there are those of us who do believe this." His young gaze strafed the room. "A number of us have suspected for years Aen's kidnapping and the death of the previous Vassal were related. Once Aen was returned and she fell ill, we were sure of it. And through Torren's help and that of others, we've been able to identify some of those involved and what is needed to cure Aen and wake her from her sleep."

The room exploded with questions.

"Yet more proof of why they shouldn't have been allowed to speak." Valerian's deep voice boomed across the room, his hands raised for silence. "Such blatant fabrications—"

"Are they fabrications, Valerian? Can you be so sure? And will you have me silenced as well because I believe in them?"

Stunned silence gripped the room at Mallean's calm words.

"I do believe our last Vassal was murdered. No one disputes the fact Aen was kidnapped. And her sleep is anything but natural."

"But...But how could Landers do all those things?" someone asked. Confused looks flooded the audience.

"How? Look before you." A half-hysterical voice rang out from the back. "One is standing right there. And what he's done wasn't enough. Now he's killed another."

Rather than get angry at the accusation, Mallean laughed. There was an edge of sorrow in the sound. "We would like it to be so simple, would we not? Better to blame the Landers, or even one of us who looks similar to them, than to see the uglier truth—that a Chosen killed the Vassal, that a Chosen stole Aen. That a Chosen has poisoned her now that she's returned."

A roar rolled through the room, filled with protests, denials, threats and more. The guards gathered close around Micca and Torren, but this time it was to protect them.

"Silence. *Silence!*" Valerian tried to quiet down the crowd.

Shock, distrust and fear were in most of the faces as the councilors slowly regained control of themselves.

"This session was not convened for such matters," Valerian declared. "We are here to determine whether these men are responsible for the death of Elon. We must strive to remain on course. These fantastical allegations can be looked at another time." He paced before them. "And to move this meeting toward its inevitable conclusion, I wish to now show you the one piece of evidence that, more than any other, proves the guilt of those before us."

From inside his cloak, Valerian removed a dagger and held it high for all to see.

"This is the weapon responsible for the death of Elon. This is the instrument used to destroy a life sacred to El. As you can see, it is a Lander's weapon."

Torren frowned. He'd been outmaneuvered yet again. The dagger was his. It was the one he had left in his pack.

"Will you deny this is yours?" Valerian's eyes shone with glee.

He stood up straight, not letting any of his trepidation show on his face. "It is. But I did not murder Elon. The dagger was stored in Aen's house."

"Lies, all lies!" screamed someone from the benches.

"He does not lie!" Micca looked worriedly from the knife to those viewing it. "Torren hasn't been here. He's been helping to find a cure for Aen."

"Yes, he hasn't been here. Escaped after his deed, helped by his accomplice" was Valerian's cold reply. "As I have seen the body myself, I can well testify it'd been rotting for at least two days. I will never forget the smell of it."

Suddenly, a boom reverberated through the council room from the chamber's doors.

"What? Who would dare...?"

The guards' commander and two others rushed up the aisle to the doors at Valerian's signal. The heavy booming reverberated once again before the bar could be lowered, and the door cracked open.

"Move aside and let us pass!"

Torren glanced toward the doors, startled as he recognized Rux's voice. He'd forgotten all about the ambassador and the twins.

"Uncle." Micca's face filled with unexpected hope.

The doors were swung wide as a gasp echoed down to

them. Everyone rose to their feet to try and see what had caused it, cutting off any chance of a view for those on the central floor. More sounds of surprise wove through the crowd as those close to the doors parted to form an avenue to the center of the room.

Rux came into view, his wings spread wide to open the way. His eyes gleamed with triumph, but his expression was somber. He nodded toward Mallean, a half-smile flashing momentarily on his face. He then made his way straight to where Micca and Torren were tethered.

Torren was able to glimpse the twins, both carrying a bundled form on their linked arms. His breath caught as he realized who it must to be.

"Larana..."

"Aen!" Micca's face broke into a grand smile.

At the sound of his voice, the form moved, a small, callused hand sneaking out to pull back the cloth surrounding her head. Larana's face was very pale, but her eyes shone brightly.

Torren's gaze met hers, her hair clip hot against his skin. She smiled warmly, tears rising in her eyes. He shocked himself by doing much the same.

"It's the Vassal! Aen's been returned to us."

Now that her face could be seen, the councilors crowded forward, some with amazement on their faces, others openly weeping.

"How?" Valerian stared, an incredulous look on his face.

"El has finally answered our prayers." This pronouncement was repeated throughout the room.

A bench was cleared, and the twins set Larana on it. They stood beside her protectively. Tyleen brought up the rear and settled behind her.

"Yes, El answered our prayers by sending Torren to us." Rux stated this loudly for all to hear. "He is the one who figured out what was wrong with her. He is the one who sent the medicine that cured her."

Questions erupted in a deluge. Torren didn't hear them; all he could do was stare at Larana. She was awake. She wouldn't sleep forever because of him.

Zelene, tears running down her face in happiness, undid Torren's and Micca's bonds.

"This is a trick!" Valerian's voice boomed above all the cries of joy and disbelief. "This man is the one who poisoned Aen, the one who killed Elon and who has now miraculously come up with an antidote to further attempt to place us in his debt. He must not be allowed to further contaminate our way of life."

For the first time, most of the comments were in opposition to his sentiments. Valerian didn't seem to notice this, turning from them to glare at Torren.

"You will not destroy us."

"Destroy you? I'm not the one trying to do that," he said quietly. "You're the one leading the Chosen on a path of self-destruction."

Valerian's face flushed. "You're a fool. You of them all should be the one to understand, but you've been blinded even more than the rest."

Torren shook his head. "No, I'm the one who sees. For now I know it was you. You are the one behind all of this."

The voices around them dropped to silence at this declaration.

Valerian laughed without humor. "More Lander tricks, I see."

"No. Just facts." Torren made sure to place himself between the councilor and Larana. "I was but a child when the previous Vassal died, but you were already grown. Records show you have spent time on the surface—a lot of time over the years—making contracts with Landers, learning their ways. This gave you plenty of opportunity to hire those who would take Aen. And once you stole her, to have them give her into the care of a Lander couple. That is, until you were ready to have these same men, years later, bring her back to you, probably with proof the original crime was committed by Landers.

"For only then would your plot come to true fruition—to turn the Chosen totally against them.

"So, when things went wrong and Aen came back on her own, you took steps to get rid of her without killing her so a new Vassal wouldn't be born. In this way, you could return to your original plan without interference."

"What lies!" Valerian thundered. "Vain attempts to turn the guilt from yourself and escape the penalty for Elon's death."

Torren slowly shook his head. "But you forget, councilor, I didn't kill Elon. You did."

"And why, by all that is El, would I do such a thing?" Valerian's composed smile was feral.

"Probably because he knew too much. I would wager from your comments he was your liaison to the Black Lords. He may very well have been the one to bring you the poison, not knowing what it was. You lied to him about me, as you did about many other things.

"But you never expected him to take matters into his own hands. You didn't anticipate he would confront me about my true origins. He was actually shocked when he discovered your allegations about me weren't true. He questioned you about it later. He wanted to know your motives. He wondered what else you'd seen fit to lie to him about.

"So, since he would no longer conform to your will and could expose you, you killed him."

"Preposterous." Valerian could barely spit the word out, his face livid.

"There's one way to find out what is true." Larana's voice brought them up short. "If I touch each of you, I can tell beyond a doubt if one of you did the deed or not."

"That's right," Icos proclaimed. "Touched as she is by El, she will sense the truth. It is how it has always been done."

"Yes," Mallean concurred, "since Aen is back among us, we should take care of this as we did before she was taken."

Torren frowned at the idea, though he held no doubts as to his innocence. He knew Larana possessed strange abilities, but he didn't believe they stemmed from a god. Had he ever heard of the Vassal holding such powers? He couldn't remember.

"Yes, yes, let Aen decide." More and more of the councilors voted their assent with relief. Others seemed more eager for it as proof El was truly back amongst them.

Larana motioned for Torren to come near. He noticed her eyes were as blue and innocent as ever, yet at the same time appeared filled with something more. Not daring to hesitate, he knelt before her.

A brief smile touched her lips as she gazed at him, and then she placed her hand over his.

He felt the strange sensations that had accompanied the other instances of her touch. He could feel her weakened condition but also the strength inside her, the joy she felt at seeing him again and something else he couldn't quite identify.

Staring into his eyes, Aen spoke. "Torren, son of Lar, did you kill Elon?"

"No."

"Have you been responsible for any of the calamities which have befallen the Chosen?"

"No."

Larana's smile was bright. "He speaks the truth."

"Yes!" Zelene wept in rapturous joy, her son vindicated.

Larana released his hand, and he stepped to the side. Micca squeezed his arm, grinning. Rux looked pleased.

Larana turned to Valerian, who stood with his wings wrapped tightly about him. "Now it is your turn to be judged."

He didn't move. He stared at her, complex emotions sweeping across his face. The guards who'd so callously dragged Torren and Micca before the council stepped forward to encircle him.

"Come," she said. "Let El see into your heart."

A choking sound issued from the councilor's throat. A moment later, his wings swept back as he lunged forward, the dagger he'd used to incriminate Torren in Elon's murder in his hand.

"This is all your fault!"

Torren stepped into the councilor's path and was able to catch his arm before it reached her. His angle was bad, and he couldn't immediately get the upper hand. Struggling, the two went down. Screams filled the air.

His wings flapping madly, Valerian pressed against Torren's strength, trying to bring the dagger down on his throat. Grimacing, the hard stone floor against his back, he struggled to hold Valerian's arm back as the blade came within less than a hand-span from his flesh.

Finally seeing an opening, he brought his knee up,

driving it into Valerian's side and throwing the councilor off-balance. This was all he needed. Shifting his grip, he twisted Valerian's hand until with a yelp he was forced to release the knife. Torren then threw him to the ground.

Valerian was yanked off the floor by the guards before he could recover.

"My son, are you all right?" Zelene helped Torren sit up, looking frightened. "Did he hurt you?"

"No, I'm fine." He got to his feet, keeping his eyes locked on Valerian, just in case. "Are you?"

She nodded and leaned against him, wrapping her arms about him. "You saved her. You saved her."

He held his mother, glancing at Larana then once more at the councilor. Valerian's hate-filled eyes met his.

"Bring him to me." The command in Larana's voice was unmistakable. Torren stared at her in astonishment, wondering where it'd come from.

Held by four others, Valerian was brought forward despite his vain struggle to get away. Pulling on his arms while pushing on his shoulders, they forced the councilor to kneel before her. Defiance and fear flared in Valerian's eyes as Larana gazed at him.

Gently, reverently, she took his face in her hands, locking her gaze to his.

"No!" Valerian jerked, trying to rise, to escape; but the four Flyers held him still. Torren moved closer.

Valerian quieted as everyone waited breathlessly for what Aen would say or do. Torren felt the first stirrings of concern as Larana sat incredibly still, staring into Valerian's eyes for what seemed an interminable amount of time.

They were all surprised when tears formed in her eyes and rushed down her cheeks as Valerian suddenly howled in tormented misery. Larana dropped her hands from his face, her eyes not leaving his.

"You should have told." She sat back, seeming to sink into her wrappings. "You should have told."

Tyleen reached from behind to steady her. Valerian hung his head, looking at no one.

"Aen!" Everyone's attention swung to the Vassal at Tyleen's cry. Larana was slumped against her.

It was all the distraction Valerian needed, He twisted in his captors' grip and reached for her, a snarl on his face.

Torren once more threw himself between them.

"Stop him!"

Valerian hit him hard on the side of the head, making it ring. The councilor then took to the air, a number of others doing the same to try and stop him.

"You can't leave this room, Valerian, so don't make this any harder than it has to be." Rux was clearly prepared to go after him himself if necessary.

Valerian glanced around, his features crumbling as he saw all had turned against him. "What she saw is a lie. It never happened. All I ever did was promote our race—for the glory of the Chosen. I only wanted to protect us from the vermin, the vermin that need to be destroyed."

"That is not El's way," Zelene told him. "And His way is our way."

"No!" Valerian confronted her, full of desperation. "He made us better than them. Superior to them. But they keep trying to taint us, to bring us down to their level out of jealousy. They can't be allowed to do this.

"Don't you realize our dependency on them makes us weak? We are superior. We should be the ones in control, the ones showing them the way. Why must we suffer at their whims when we are so much more deserving?"

Torren watched the councilor weave back and forth, though occasionally he sent a concerned glance in Larana's direction. The guards were unobtrusively closing in on Valerian. He would soon have nowhere to go.

"We cannot enslave them, Valerian." This came from Mallean. "Though we at times forget, our people were once Landers. It was Landers who helped El in His time of need. It was Landers He took and made His own." Her voice was filled with sadness and pity. "As you've shown us, we are still like them in every way. We even have the ability to kill."

"It doesn't matter. Can't you see that? None of it does. We're good and they're not and so they must be destroyed." Valerian slowly floated to the floor. "Look at what they did to *him*. How they robbed *him*." He pointed at Torren. "It is one thing to be infirm, or to lose them due to accident, but to be maimed for sport? They would do to you what they did to him without hesitation."

Torren shook his head. "No, not all of them are that way. I didn't want to admit it for a long time, but it's true. You can't destroy them all just because a few are evil." His voice turned hard. "Besides, it was you, a Chosen, who ordered my father and the others killed. This was done to me because of you."

Valerian flinched at the words.

"So, not only is the blood of Elon on your hands, so is the blood of my father and his friends. Any and all who've had their lives destroyed because of the Black Lords."

Here, finally—the true cause of his misery. He felt his anger rising toward rage. His fists clenched and unclenched at his sides. It hadn't been the god of the Chosen, it hadn't been because of Aen. It was because of the delusions of this twisted man.

"You are responsible for everything."

"Torren, no!"

He stopped, only then realizing he'd strode forward, violence on his mind. How did Larana know? He glanced back at her, his hatred barely held in check now that the true cause of his misery was so close at hand.

She was sitting up, supported by Tyleen from behind. She looked sad, and very weak.

"He is not well, Torren. When he was very young, he left the embassy, some evil men found him. They—"

"Do not speak of it." Valerian's screech echoed through the room. "You will not speak of it."

The guards made their move.

He saw them coming and tried lifting into the air. Several collided with him, and they all fell to the ground in a heap. A snapping sound filled the room followed by a howl of pain. One of Valerian's wings now hung askew.

Torren saw something glitter on the ground, and with a sinking feeling realized no one had picked up his dagger from the previous struggle. Almost as if reading his mind, Valerian spotted it and swept it up before anyone could stop him. He gashed one of the guards on the arm and another in the back. The councilor's eyes grew wide with fear and were clouded with pain.

"Not again, not again."

Torren and other Chosen now rushed forward, trying to capture the blade. As they closed in from all directions, Valerian suddenly screamed in abject terror. One, two, three of the Flyers were slightly wounded before they became too much for him.

The knife was taken from his hand, his arms restrained. Valerian stared at those around him, his eyes filled with madness.

"El does not exist!" His voice was shrill. Several of the guards backed away from him at the blasphemy. "I tried to always keep this from you, to help you, guide you, but he does not exist! And I proved it. For if El did exist, he wouldn't let his Vassal be hurt, he wouldn't allow any of us to be hurt.

"But did he save him? Did he show his anger across the sky and smite me? No!" Spittle sprayed from his mouth.

Several of the councilors turned away, holding their hands over their ears. More simply stared at him, stunned.

"If Aen could be killed, if his replacement could be kidnapped, then anything could happen to us! We are too complacent, while the grubs outnumber us thousands to one! How long will it be before they figure out a way to overrun us, to reach our cities, to hunt us down and eat our flesh? I did what I had to do to assure our survival! I tried to be for you the god we never had!"

"Blasphemer!" One of the guards struck him, driving him to his knees. Valerian cackled with insanity as tears coursed down his face to stain the floor beneath him. Another guard and another raised their hands to inflict further blows.

"Stop!" From out of nowhere, Larana was suddenly at Valerian's side, protecting him with her body as she swayed on her feet. "This is not the way. This is not El's way."

Valerian cackled again, his body slumping to the floor. "He doesn't exist, he doesn't exist."

He looked nothing like the strong councilor all had relied on for so long.

Torren stared at the source of his misery and found he

felt almost nothing. In his own way, Valerian, too, had been a victim. He would never forgive him, but he would pity him. Such was the end of the man who'd robbed him of his wings, of his life and the life of his father.

Many of those in the room turned away from the broken man. Some wandered aimlessly about, as if too shocked to know what to do. Others stared at Valerian with disgust or anger. He doubted anyone here would survive untouched by what they had witnessed this day.

"Torren." Larana took a shaky step in his direction, her arms extended toward him, tears running unhindered down her face. Without thought, he closed the distance between them and took her in his arms. She buried her face in his chest, her thin frame wracked with sobs.

"We need to get her out of here."

Nodding, and looking grateful for something useful to do, Rux took command. "You three, go open the doors. You others take Valerian to the healer's aerie. Do not let him out of your sight. Everyone else, move outside to the fresh air."

Torren scooped Larana up in his arms, and with the twins helping him, got her out of the council room. Once outside, he didn't stop. Ignoring everyone and everything, he took Larana home.

By the time he set her down on her bed, she'd quieted, though she still clung to him.

"Please don't leave me, Torren. I need you."

He hesitated, hoping she only meant at this moment. "Don't worry yourself. I'll sit here beside you. Just rest."

Larana inched over, still holding on to him, and made room. He sat down, his back against the wall; and she nestled up against him, dark circles very prominent beneath her eyes.

"I didn't know men could hurt children that way," she said in a soft whisper a few moments later. "He hated them so much…"

Torren caressed her cheek, moving an errant curl of hair from her eyes. "Just rest. It won't help to think about these things right now."

Her dark eyes rose to meet his, her sadness suddenly replaced with something bright. "You did all this, you know. He knew you would."

He frowned, not sure what she was talking about.

"You don't believe, but He has always been with you. He couldn't break the rules set down by the First Mother, so instead He bent them and spoke to you in the only way open to him." Larana's eyes closed, her voice growing softer still. "He showed it all to me. It was why I didn't know Him when we found each other. Why I didn't have the knowledge everyone expected. *You* were the Vassal then, Torren, not me."

"What?" He shook his head, sure he'd misheard.

But Larana didn't respond. Her breaths were deepening, her body relaxing as she fell into a deep sleep. He recalled the words of the First Mother's priestess, telling him to "fix the Vassals"—plural.

No, he must have heard it wrong. No gods existed, not even theirs; in this, he agreed with Valerian. Not once had he seen proof of any of them. In the last weeks, he'd witnessed a number of strange things—Larana's unusual gifts, the woman Mala, even his own almost prophetic dreams. Things other than godly interference could have caused all of it. Right? Gods were an invention of society. Something to help unify people and give them hope, something to believe in. His past had burned out of him any real belief in such illusions.

Yet suddenly, a few misheard words from this girl had him questioning things he'd long thought settled. It was one more reason why he needed to leave.

Tomorrow. He would return to his old life tomorrow. His mother, Mallean, Micca—the whole of the Chosen population would take care of Larana. And with Valerian exposed, she would be safe. Besides, there were scores to be settled with the Black Lords. He was sure Von Duren wouldn't mind his assistance in that regard. He might even get involved in trying to explain things to the Galts. Living up there a while might do for a nice change.

Yes, that's what he would do.

But even in his own mind he didn't seem as sure this was the right course for him anymore...

# CHAPTER 27

Torren was in the air. The ground was distant and moved beneath him. The wind wove through his hair. Around him was the immensity of the blue, open sky.

A thrill long-forgotten filled him as he turned to face upwards and flapping his wings, feeling the muscles in his back responding as he soared to even greater heights. This was El's gift, His gift to those land-bound, as He'd become when the First Mother allowed Him to become human. This was what El had done for the Chosen.

Gone was the fear he had felt when Micca lifted him into the air, gone was any trace of worry he'd fall. This was what he was; this was what he'd been meant to be. He wrapped himself in his glorious white wings and dove, feeling the truth of this in every fiber of his being.

———

He slowly opened his eyes, the feeling of belonging, of being, still wrapped about him. Then, like a flame suddenly doused with water, he remembered who and what he was, and the reality soured the feeling into bitter loss. He was no longer a Chosen. He could no longer fly. The dream had been of something that could never be.

Disgusted his mind could call up such fancy, he realized for the first time Larana was no longer with him. He sat just as he had fallen asleep; but not only was she gone, someone had taken the time to tuck a blanket around him. He was amazed this alone hadn't wakened him. It would have been better than to dream.

He was still for a moment, listening for sounds of the

girl, to see if she was still nearby. Hearing nothing, he realized he truly was alone. It puzzled him, especially after her earlier insistence that he stay.

Getting up off the bed, he pulled the thin curtain demarcating the room aside. No one was there. His heart beat faster—surely, nothing had happened to her?

Torren hurried through the linium, still not seeing anyone. As he reached the doorway into the sorium, though, he slowed. Soft music drifted to his ears, as well as a number of voices mixing sweetly with it. From his current vantage point, he noticed all the outer curtains to the sorium were pulled back, opening the area to the outside. The music and voices were coming from there.

Taking a closer look, he saw people crowded up on the steps, making quiet lines, each in turn bowing before taking their leave of the small figure seated on a mound of cushions. Tears and smiles of happiness and thanks glistened on the faces of those paying homage to El's Vassal.

The twins were there as well, fully dressed in their armor, their smiles as bright as the rest. Micca stood behind them, also in full armor; but his plumed helmet rested in the crook of his arm at his side. Mallean, Zelene, Icos and a few of the other councilors rested close by, their faces worn yet eased somewhat from the strain of the last few weeks.

Torren felt relief run through him at the sight, even as a touch of melancholy colored his thoughts. This time it was truly finished. What his father had set out to do all those years ago was done. Larana was safe, loved by the Chosen. As for him, he'd been allowed to do it.

*Allowed?* The thought was strange, but he left it alone. It wasn't the time to ponder such things.

Still, he was glad. The wounds of the Chosen could now heal. His own were mostly sealed—they would not bleed anymore, and with time the scars would fade. He knew he wouldn't dream of the deaths of the others again. Their souls would now rest.

He was as close to feeling at peace as he had ever come. It startled him as, for the first time in his life, he realized he felt a little hope for his own future.

Larana's laughter drifted toward him, eliciting a faint

smile. Yes, hope. Nothing would be as it'd been meant to be, but perhaps he could learn to live with his people again, be one of them despite his handicap. He'd faced his worst fears and survived them. Found support from unexpected quarters in both worlds at different times.

He slowly shook his head, amazed by his thoughts. How had all this changed him? Could one night's rest truly make such a difference? Last night, he'd still been so sure he would leave, never to return, yet now...

He turned, meaning to depart by another way so as not to disturb the others, confused by his sudden uncertainty. He would go see his aunt, have her come get his mother for him. There were things he needed to tell her he might be able to now; and he wanted to explain he wouldn't be leaving permanently this time, just for a while, despite his previous plans. He didn't want to make the same mistake he had with his foster parents. Didn't want to bring her pain she didn't deserve.

Maybe she could then help arrange for him to be able to spend a few minutes with Larana alone so he could take a proper leave. Rux had some work to do, and if he volunteered to help him instead of Von Duren he'd be made to deal with his own kind. That would be hope's first step. He wasn't sure what he would do for the ambassador, but surely he could find something. And by the time the island came around again, perhaps then...

"Torren!"

He flinched, having thought he'd remain unnoticed. He'd had more than enough of being the center of attention. It didn't suit him. Micca started toward him, a number of others noticing his presence with the shout.

He was half-tempted to ignore him and leave, but by then Larana was beaming at him. Torren made himself walk into the sorium.

Micca gave him a warm grin and a greeting.

"I was starting to think you'd sleep another day away." A touch of color flashed in his cheeks. "I wanted to go wake you, but Aen wouldn't hear of it."

Impulsively, he reached out and hugged him. Torren was too startled to protest.

"Thank you. Thank you." Micca pulled away, his eyes

bright. "I've been wanting to do this all day. You brought her back to us not once but twice. And who knows what worse things would have happened without you."

He wasn't sure what to do with all the heartfelt gratitude. "I didn't do anything you wouldn't have done."

"He's always been a modest one."

Torren almost fell forward as he was slapped hard on the back.

"Sal?" He glanced behind him.

His friend grinned. "None other."

"But...how?"

"Dom Rux came down and filled us in on what happened." His grin grew wider. "Then he handed invitations, all proper-like, to me, the governor and Von Duren to come on up here and meet the Vassal. Wasn't passing up the opportunity of a lifetime, I can tell you. So, here we are." He slapped him heartily on the back again.

Torren stared at him, his mind reeling. Landers on Chosen soil? As far as he knew, this had never happened before, except perhaps back in the early years.

"Torren!"

He was almost knocked over as Larana crashed into him, wrapping her arms about him.

"Aen, please, you're still weak. You should not overexert yourself." Tyleen was close behind her, her pretty face flustered. She eyed him and Sal with some misgivings.

Larana paid her no heed, looking up happily into Torren's face. "He told you, didn't He? That He's giving you a gift?"

Torren frowned, not understanding. A number of the others were now gathered around him. What were they all smiling for?

Larana let him go, but before he could move away she took his hands in hers. Her overwhelming joy came across loud and clear. "He loves you, He loves all of us. And you saved us as well as the First Mother's children. It's why He is now able to give you this gift. A gift that will mean much to you—and to everyone."

"No. I don't..." Not sure what he was refusing, and not liking who it was Larana was implying it came from, he fervently wished he'd been able to get away unnoticed.

Why was she saying these things? Nothing would happen. The recompense would be in money or position, and he didn't want either of those things.

"He knows what you desire. He wants to give it to you. To help you be whole again."

Torren's gaze locked with Larana's, the utter conviction in her voice and the emotions coming from her touch too intense to ignore. He gasped suddenly as a concentrated itching sensation flashed just below each of his shoulder blades.

"Please everyone, get back!" She released him, her smile even brighter than before and waved for those around him to step away.

Suddenly afraid, though he wasn't sure of what or why, Torren meant to turn away; but his knees grew abruptly weak and brought him to the ground.

"Wha—"

Music filled his mind, and from there flooded the rest of his body. His breathing grew ragged; but he was barely aware of it, the beauty of the music distancing him from his physical self.

"What's wrong with him?" Sal's concern was obvious.

"Nothing is wrong," was Larana's happy reply. "Things are finally being made right."

He heard them, but it wasas if from a distance. Braced on hands and knees, his body jerked, but there was no pain; it was diffused by the music.

"Micca, help me," Larana said in an excited voice. "We have to cut his shirt."

Unable to do anything to stop them, Torren felt a shadow of shame rise, for his scars would now be exposed, even though he'd thought it didn't matter as much as before. Like a gentle lover, however, the music caressed him and took the shame away. He felt cool air touch his skin as his shirt was carefully cut away from his body.

The music swelled in a crescendo, and it swept him up with it. His back spasmed as he felt a great pressure fill him, rising with the music. All at once, the pressure swooshed out from him through his back. As he reeled with relief and dizziness, gasps rang all around him, harmonizing with the music still ringing in his mind.

Then, suddenly, it stopped. Weak, not sure of what had just occurred, Torren closed his eyes and let himself drop. He didn't hit the floor. Hands clamped on to him and pulled him up. Before he could panic, cushions were placed beneath him, and he was helped to sit. He felt strangely unbalanced.

Larana stood before him, holding a cup to his lips. "They're beautiful, Torren, beautiful!"

Drinking the strong wine, he blinked at her with little understanding.

"Look."

Attempting to do as she bid, he followed her pointing finger over his shoulder.

"It's His gift to you. His gift. So you can believe again. So everyone can."

Torren's eyes widened as he forgot everyone and everything but what met his gaze. They arched behind him—gold, not the usual white—and wondrous. Disbelieving, he scanned one, extending it to make it easier to see.

The feathers appeared soft and downy like a newborn bird's. In infinite wonder, he reached up with a shaking hand to touch his new wings.

### END

# ABOUT THE AUTHOR

Born in 1964 in Rio Piedras, Puerto Rico, GLORIA OLIVER bounced around several states during the teenage years, finally ending up in Texas for good. Married for twenty years, she is the proud parent of a very independent daughter. She originally entered the University of Texas in Arlington to obtain an aerospace degree,but eventually moved over to the University of Texas in Dallas to gain a BA in Interdisciplinary Studies, and is currently working in the finance/accounting field. Her hobbies at present are reading, writing, watching Japanese animation, collecting music and translating Japanese comics.

# ABOUT THE ARTISTS

MARTINE JARDIN has been an artist since she was very small. Her mother guarantees she was born holding a pencil, which for a while, as a toddler, she nicknamed "Zessie."

She won several art competitions with her drawings as a child, ventured into charcoal, watercolors and oils later in life and about twelve years ago started creating digital art.

Since then, she's created hundreds of book covers for Zumaya Publications and eXtasy Books, among others. She welcomes visitors to her website: www.martinejardin.com.

www.ingramcontent.com/pod-product-compliance
Lightning Source LLC
Chambersburg PA
CBHW022054210326
41519CB00054B/350